Discourse and Creativity

Discourse and Creativity

Edited by Rodney H. Jones

Routledge
Taylor & Francis Group
LONDON AND NEW YORK

First published 2012 by Pearson Education Limited

Published 2014 by Routledge
2 Park Square, Milton Park, Abingdon, Oxon OX14 4RN
711 Third Avenue, New York, NY 10017, USA

Routledge is an imprint of the Taylor & Francis Group, an informa business

ISBN 13: 978-1-4082-5188-1 (pbk)

British Library Cataloguing-in-Publication Data
A catalogue record for this book is available from the British Library

Typeset in 9/13.5pt Stone Serif by 35

Contents

List of figures and tables

Figures

Tables

Contributors

Vijay Bhatia is the founding President of the Asia-Pacific Association of LSP and Professional Communication. He has published on the analyses of legal, business, newspaper, and advertising genres; ESP and Professional Communication (Theory and Practice); simplification of legal and other public documents; and cross-cultural and disciplinary variation in professional discourses. His work on genre analysis, which includes two of his best-known books, *Analysing genre: Language use in professional settings* (Longman) and *Worlds of written discourse: A genre-based view* (Continuum), is widely used as a resource for research and practice in ESP, discourse and genre studies.

Stephen Bremner is an Assistant Professor in the Department of English at the City University of Hong Kong. His research interests include organisational communication, English for Specific Purposes, collaborative writing, and politeness theory.

Charles Forceville is employed in the department of Media Studies of the Universiteit van Amsterdam, where he directs the department's Research Master programme. His teaching and research are inspired by cognitivist-oriented models in linguistics (Conceptual Metaphor Theory, Relevance Theory) and film (Bordwellian approaches) and focus on the structure and rhetoric of multimodal discourse in various media and genres (advertising, documentary, animation, comics, cartoons). He is author of *Pictorial metaphor in advertising* (Routledge) and co-editor of *Multimodal metaphor* (Mouton de Gruyter).

Julia Gillen is Senior Lecturer in Digital Literacies in the Department of Linguistics and English Language at Lancaster University and co-editor of the *Journal of Early Childhood Literacy*. Among her publications are *A day in the life: An international study of two-year-old girls* (Palgrave Macmillan) and *Researching learning in virtual worlds* (Springer).

Martin Havnør has a Master's in Informatics from the University of Oslo. He has collaborated in a number of practice-based research projects in digital communication and does so through his current position at Faster Imaging, a leading mobile concern in Norway.

Rodney H. Jones is Associate Head of the Department of English at City University of Hong Kong. His research interests include youth cultures, computer mediated communication, multimodal discourse analysis, discourse and sexuality, and health communication. He is co-editor of *Discourse in action: Introducing mediated discourse analysis* (Routledge) and *Advances in discourse studies* (Routledge), and co-author with Ron and Suzanne Scollon of *Intercultural communication: A discourse approach* 3rd edition (Wiley-Blackwell) and with Graham Lock of *Noticing, exploring and practising: Functional grammar in the ESL classroom* (Palgrave).

Henry Mainsah has a PhD in Media and Communication from the University of Oslo where he lectures on media, cultural identity and globalisation. His research interests include cultural studies and design ethnography. He is currently a guest researcher in design at the Oslo School of Architecture and Design.

Andrew Morrison is Professor of Interdisciplinary Design at the Oslo School of Architecture and Design where he leads the Research by Design Centre (RXD). His latest book is *Inside multimodal composition* (Hampton).

Sigrid Norris is Associate Professor of Communication at Auckland Institute of Technology and head of the Multimodal Research Centre. Her book *Analyzing multimodal interaction: A methodological framework* (Routledge) is widely used. She is also co-editor (with Rodney Jones) of *Discourse in action: Introducing mediated discourse analysis* (Routledge).

Anne Peirson-Smith is Assistant Professor in the Department of English at City University of Hong Kong. She is co-author of *Public relations in Asia Pacific: Communicating effectively across cultures* (Wiley) and *Introduction to fashion* (Berg).

Idunn Sem is a media developer at InterMediaLab, University of Oslo. She has wide experience in practice-based research in new media, learning and design. Since her Master's studies, she has published in areas such as dance and technology, digital literacies and mobile communication.

Peter Stockwell is Professor of Literary Linguistics at the University of Nottingham. His research interests include stylistics, narratology, cognitive poetics, and the sociolinguistics of literary reading. He is author of *Cognitive poetics: An introduction* (Routledge) and *Texture – A cognitive aesthetics of reading* (Edinburgh University Press).

Joan Swann works at the Open University, UK, where she is currently Director of the Centre for Language and Communication. Her interests centre on the development of a socially oriented and contextualised approach to language and creativity. Recent publications on language and creativity include special issues of the journals *Applied Linguistics* and *Language and Literature*, and the book *The art of English: Everyday creativity* co-edited with Janet Maybin (Palgrave Macmillan).

Crispin Thurlow is Associate Professor of Communication at the University of Washington (Seattle), where he also holds an adjunct position in Linguistics. His books to date are *Computer mediated communication: Social interaction and the internet* (Sage), *Talking adolescence: Perspectives on communication in the teenage years* (Peter Lang), *Semiotic landscapes: Language, image, text* (Continuum) and *Tourism discourse: Language and global mobility* (Palgrave Macmillan).

Michael Toolan is Professor of English Language and currently Head of Department in the Department of English at the University of Birmingham, where he teaches courses in Stylistics, Language and the Law, and Narrative Analysis, and convenes the MA programme in Literary Linguistics. His books include *The stylistics of fiction* (Routledge), *Total speech* (Duke University Press), *Language in literature* (Hodder Education), *Narrative: A critical linguistic introduction* (Routledge) and *Narrative progression in the short story: A corpus stylistic approach* (Benjamins).

Theo van Leeuwen is Dean of the Faculty of Arts and Social Sciences at the University of Technology, Sydney. Previously Director of the Centre for Language and Communication at the University of Cardiff, he initially trained as a film and television writer and producer and has maintained a strong interest in the synergy between theory and practice throughout his career. His research interests include media discourse, critical discourse analysis, and multimodal communication. He is one of the founding editors of the journal *Visual Communication* (Sage) and has had ten books and 81 journal articles and book chapters published.

Publisher's Acknowledgements

We are grateful to the following for permission to reproduce copyright material:

Figures

Figure 7.1 adapted from *The Way We Think: Conceptual Blending and the Mind's Hidden Complexities*, Basic Books (Fauconnier, G. and Turner, M. 2002) p. 46, copyright © 2003 Gilles Fauconnier, Mark Turner. Reprinted by permission of Basic Books, a member of the Perseus Books Group

Text

Poetry on pages 26–7 from *The Spirit Level*, Faber and Faber (Heaney, S. 1996); Lyric on page 135 from *Fine and Mellow*, Billie Holiday, copyright Carlin America, Inc., FINE AND MELLOW Words and Music by Billy Holiday © Edward B Marks Music Company – All Rights Reserved – Used by Kind Permission of Carlin Music Corp., London NW1 8BD

Photos

(Key: b-bottom; c-centre; l-left; r-right; t-top)

Alamy Images: Classic Image / Alamy 122; **Andrea:** 154, 158; **Dr Ch.J. Forceville:** 121; **Mike Garofalo:** 184; **Image courtesy of The Advertising Archives:** 175; **iStockphoto:** 172; **Andrew Morrison:** 220, 221, 222, 223; **Schome:** 196, 197, 199, 202; **The Nestlé name and image is reproduced with kind permission of Société des Produits Nestlé S.A:** 123, 124

Cover images: *Front:* **Getty Images**

In some instances we have been unable to trace the owners of copyright material, and we would appreciate any information that would enable us to do so.

CHAPTER 1

Introduction: Discourse and Creativity

Rodney H. Jones

This collection presents a range of different perspectives on the relationship between *discourse* and *creativity*. It is divided into four sections, each focusing on a different type of discourse: The first section explores literary discourse, the second focuses on creativity in corporate and professional discourse, the third on creativity in multimodal discourse of various kinds, including advertising graphics, fine arts and music, with the final section addressing the impact of new technologies on creative texts and practices. In bringing together studies of creativity in such a wide variety of genres, media and modes from poetry to amateur skateboarding videos, and from such a variety of perspectives in discourse studies, from more traditional literary stylistics to newer approaches such as multimodal and mediated discourse analysis, this volume aims to explore the different kinds of contributions discourse analysis can make to our understanding of creative products, the social and psychological processes that go into making them, and the ways they help to shape the identities, relationships and institutions that make up our societies.

What is Creativity?

In the last two decades, the notion of 'creativity' has found its way into nearly every facet of human life, from education to management. A hundred years ago, creativity was seen primarily as the province of artists (poets, painters, composers) and of God. Nowadays, everyone is expected to be creative. A cursory search of the British National Corpus of written and spoken English finds 'creative' collocating with such diverse words as accounting, bankruptcy, competition, governance, management, manufacturing, privatisation, recreation and relationships. The last fifty years has seen a proliferation of popular books, courses, and position papers from governments and other institutions on how to make people, businesses, organisations and societies more creative.

This 'democratisation of creativity' (Maybin and Swann, 2007) is also reflected in academic research in a range of disciplines such as psychology, sociology, anthropology and linguistics, which has turned its attention to the everyday creative practices of ordinary people. In such studies, creativity is, in the words of Ron Carter (2004: 13), seen as 'not a capacity of special people but a special capacity of all people.'

Of course, not all creativity is 'created equal'. There is a qualitative difference between writing a symphony and creatively altering a recipe when one has run out of sugar. To capture this difference, Boden (2004) famously distinguished between *historical creativity* and *psychological creativity*, or, as others have called them, 'big C Creativity' and 'small c creativity'. 'Big C Creativity' refers to the creativity of world-changing works of art or scientific discoveries that alter the way people think about a certain problem or domain, whereas 'small c creativity' refers to the creativity evident in everyday problem-solving, joking and verbal play: avoiding a traffic jam or coming up with a good pick-up line at a bar. Whereas 'big C Creativity' is seen as a sign of genius, 'small c creativity' is seen as a sign of mental health, a necessary competence for getting along in the world.

The problem with these definitions is that there is a lot in between the works of Shakespeare and a well-delivered apology to one's in-laws. Many (indeed most) efforts in art and literature that aspire to the greatness of 'big C creativity' sadly miss the mark, and many everyday acts of creativity end up, sometimes unintentionally, having a major impact on the way people think and interact with one another, even if it is often in a rather limited social circle. Most of what is presented as creativity in the following chapters occupies this middle ground. There is poetry (not all of it 'great') and music and painting, but there is also advertising, corporate and public relations writing, and the creative practices of young people using digital technologies.

This problem around what counts as 'creative' and what does not exposes an even more fundamental confusion in the way we talk about creativity. When we use the word 'creativity', are we talking about a property of a particular creative product – a text or an object of art or the expression of a scientific theory – or are we describing a kind of *process*, what an individual or group of people do to come up with a creative product or inventive solution to a problem? This is to say, does creativity reside in texts (and other social artifacts) or does it reside in people?

Most studies in the humanities, in literary and art criticism, have taken an almost exclusively *product* based approach to creativity. While some have sought to contextualise creative works in their social or historical contexts or to glean from them evidence of the workings of the artist's mind, the starting point has nearly always been the text.

In the social sciences, on the other hand, particularly in psychology, scholars have been more interested in the creative process. Psychological studies of scientific creativity (see for example Simonton, 1988; Gruber and Davis, 1988) and artistic creativity (see for example Getzels and Csikszentmihalyi, 1976) have focused on mental processes and cognitive models. Some like Csikszentmihalyi and Sawyer (1995) and Runco (1990) have offered theories of the 'stages' of the creative process as it occurs in an individual's consciousness. Others have taken a more socio-cultural or interpersonal approach to creative processes, seeing them as not just taking place in the minds of the individuals but also in the interaction between individuals and their social and cultural environments. Here we could include Harrington's (1990) ecological approach, the interactionist model of Woodman and Schoenfeldt (1990), and the systems approach of Csikszentmihalyi (1990).

Thus far, however, there has not been a clearly articulated perspective which integrates approaches which focus on the creative properties of products (by which we mean primarily 'texts', whether they be verbal, visual or expressed in some other semiotic mode) and the processes through which they come into being. Not only have product based approaches not adequately addressed issues of production and consumption, but process based approaches – which have typically proceeded by examining the practices of 'creative individuals' such as renowned artists and scientists – have been less effective in clearly identifying the concrete features in these individuals' achievements which make them creative. Even within the process approach, there remains a gap between those who take a more cognitive or intra-psychic perspective and those who take a more socio-cultural or interpersonal perspective (John-Steiner, 1992).

It is the premise of this volume that discourse analysis, particularly as it has developed in the past thirty years through contact with other disciplines like cultural studies, cognitive psychology, sociology and anthropology, can make a significant contribution to bridging these gaps. Nearly all of the chapters in this book deal explicitly with the creative processes that go into the production and interpretation of discourse, sometimes focusing more on cognitive processes, as in the chapters by Stockwell and Forceville, and sometimes more on social processes, as in the chapters by Swann and Jones. At the same time, all of them enter this exploration of process through the analysis of creative products – discourse – and it is in the concrete features of discourse that evidence for these social and cognitive processes is found. Moreover, while some of the scholars included here emphasise the psychological aspects of these processes and some the social aspects, discourse itself serves as a link between the two, the site at which is played out the eternal tension between what the individual wishes to think or do or express and what his or her society or culture deems appropriate or meaningful or 'creative'.

'Language and Creativity' vs 'Discourse and Creativity'

There has been considerable interest over the years in various sub-fields of linguistics in the notion of creativity. It might, in fact, be argued that creativity is at the very core of language itself, the 'essential property' of which is, according to Chomsky (1965: 6), 'that it provides the means for expressing indefinitely many thoughts and for reacting appropriately in an indefinite range of new situations.'

In the areas of applied linguistics and sociolinguistics, interest in creativity has led scholars in two distinct directions, some focusing on the application of linguistic principles to the analysis of texts that are *a priori* deemed 'creative' such as literary works and advertising slogans, and others focusing more on the creative and playful features of everyday language.

Scholars who take literature as their objects a study of literature include literary stylisticians such as Fowler (1996), Leech and Short (1981), Widdowson (1975) and Toolan (1998) who apply the tools of linguistics to the analysis of literary language. While some working in this tradition have endeavoured to focus on aspects of language use normally associated with 'discourse' such as pragmatics (Black, 2006), speech acts (Pratt, 1977), interpersonal politeness (Magnusson, 1999), conversational structures (Norrick, 2000), and schema (Cook, 1994), most work in this area is primarily *product based*, defining creativity as a function of 'patterns of formal features' and 'linguistic idiosyncrasies of particular texts' (Cook, 1998: 205) rather than as a function of the processes that go into making those texts or how those texts are used to take actions in broader socio-cultural contexts.

Approaches which focus less on traditional 'creative texts' and more on the creativity of everyday language are perhaps best represented by the work of Ron Carter who, in his 2004 book *Language and creativity: The art of everyday talk* and elsewhere (Carter, 1999; Carter and McCarthy, 2004) argues that features associated with literary texts like word play, rhyme, metaphor, simile, hyperbole, understatement, irony, repetition and parallelism are actually common features in the everyday spoken English of ordinary people. The hard and fast distinction between literary and non-literary language is, he contends, artificial and unhelpful; literariness is more usefully seen as a 'cline' from, to use the terminology discussed above, the 'small c creativity' of commonplace talk to the 'big C Creativity' of the literary canon. Other researchers working in the same vein include Cook (2000), Crystal (1998) and Maybin and Swann (2006, 2007).

Like literary stylistics, linguistic approaches to everyday creativity have also made use of principles from discourse analysis. Carter, for example, addresses not just the literary features of everyday talk but also the communicative functions of these features in different kinds of social contexts and in different forms of social interaction. On the whole, however, most work in this tradition is also primarily *product* oriented, concerning itself almost exclusively with 'poetic language', in the sense that Jakobson (1960: 356) meant the term as a 'focus on the message for its own sake' rather than on the role of the message in broader social processes. Even when they take socio-pragmatic aspects of language use into account, researchers in this paradigm tend to focus on the social functions of *creative language* (by which they usually mean 'literary-like' language) rather than the function of language (of all kinds) in performing *creative acts*.

How, then, does the 'discourse and creativity' approach represented in this book differ from the approaches described above? To answer this question it is necessary first to understand what we mean by discourse. While all of the authors in this book might answer that question slightly differently, most definitions of discourse in the context of applied linguistics and sociolinguistics draw on three broad conceptualisations of language: language beyond the level of the sentence or clause; language in use; and language as part of a broader range of social practices associated with power and the social construction of knowledge. It is important to stress that these three conceptualisations of language are not so much separate and mutually exclusive 'definitions of discourse' as they are different aspects of the same phenomenon, none of which can be properly understood without reference to the others. Nearly all contemporary approaches to discourse take all three of these aspects into account, though they might focus more on one or another of them.

The first conceptualisation – language beyond the sentence – can be traced back to the linguist Zellig Harris (1952), who in the early fifties used the term 'discourse' to describe the next level in an analytical hierarchy of morphemes, clauses and sentences. What Harris proposed was a method of analysing language beyond the sentence by attending to the distribution and combination of various linguistic features throughout longer stretches of text. This approach, however, is not just an extension of the Russian formalists' search for intra-textual regularities. Even in Harris's early formulation, patterns of linguistic features beyond the clause need to be further related to patterns of behaviour beyond the text itself. In his seminal 1952 paper he proposes 'discourse analysis' as a means of addressing two interrelated problems, the first arising from the fact that most models of descriptive linguistics stop at the level of the sentence, and the second arising from the need to correlate 'culture' and language, that is, to understand the connection between linguistic and non-linguistic behaviour.

The implication of a view of discourse as 'language beyond the sentence' for a 'discourse and creativity' approach is that in such an approach creativity is never seen as a matter of isolated instances of 'poetic' language, but rather as a matter of how all the features of a text, poetic or not, work together to form an effective whole, and further, how this whole interacts with the social context in which it is situated. In other words, puns, metaphors, or instances of rhyme or parallelism are not considered creative in themselves but rather are seen as creative insofar as they fit into larger patterns of structure and meaning.

This search for patterns in texts is, of course, not unique to discourse analysis. It is also central to literary stylistics in the more traditional sense. This practice of pattern seeking, of relating smaller parts to larger wholes, however, is the necessary starting point for a 'discourse and creativity' approach and for all of the chapters in this volume. It is fitting, then, that the book begins with Michael Toolan's treatment of repetition in poetry, a treatment that illustrates the attention to patterning so central to the conceptualisation of discourse we are developing in this book while at the same time paying tribute to traditional stylistics.

Implicit in this analytical stance towards creativity is also the notion that underpinning the creative process itself is the ability to recognise and exploit patterns in our experience of the world and in the semiotic systems within which we work. Bohm (1998), for example, in his treatment of scientific creativity, defines the creative process as one of perceiving new orders of relationships in old structures and of linking previously unrelated ideas, concepts or elements into new patterns. From this perspective, the relationship of patterning to creativity is double-edged. On the one hand creativity involves understanding and being able to exploit old patterns, structures and rules, and on the other hand it involves breaking out of old patterns and coming up with new ones. As Thurlow reminds us in his chapter, 'creative practice always emerges out of the dialectical tension between fixity and mobility, constraint and freedom, convention and innovation, stricture and defiance, orthodoxy and heterodoxy, and, in the case of language, between "grammar" and "poetry".'

The second conceptualisation of language, that of 'language in use,' is most commonly associated with approaches to discourse which examine, as Austin (1962) famously put it, how we 'do things with words.' Approaches such as pragmatics, conversation analysis, interactional sociolinguistics, and Austin's speech act theory all see discourse itself as a kind of social action and explore how people use it to both *make sense of* and to *alter* the circumstances of their social and material worlds. More recent approaches to discourse such as mediated discourse analysis (Jones this volume, Jones *et al.*, this volume, Morrison *et al.*, this volume, Norris and Jones, 2005), and multimodal interaction analysis (see Norris, 2004 and this volume) influenced by the work of Soviet psychologist Lev Vygotsky (1978), have gone even further in privileging social action as the unit of analysis, considering language as only one of a host of possible 'meditational means' which people use to take action in the world.

The implications of this view of discourse for a 'discourse and creativity' approach is that creativity is seen as residing not just in language itself but in the actions people take with language. There may, therefore, be nothing intrinsically 'creative' or 'poetic' about a piece of language. What may be 'creative', rather, may have more to do with the strategic way it is used to solve a problem, alter a situation or realign a set of social relationships. This view of discourse, in other words, takes us beyond the analysis of creative *products* to the analysis of the creative *processes* associated with them. These processes include not just the creative ways discourse is deployed to take situated social action, but also the processes through which creative texts are produced and interpreted, processes that often involve complex chains of social actions negotiated among diverse sets of social actors (see Jones *et al.*, this volume) using a range of different meditational means (see Morrison *et al.*, this volume).

Fairclough (1992) refers to these complex chains of action as 'discourse processes', which he defines as the sociocognitive processes by which the producers of texts draw upon and transform past conventions and prior texts to create new meanings, and the consumers of texts appropriate and adapt these meanings based on their past understandings and experiences and their present circumstances. And so again, the tension between the old and the new, the borrowed and the original, the conventional and the subversive arises at the centre of a discourse analytical approach to creativity.

All of the chapters in this book address in some way 'discourse processes' or, as Norris and Jones have called them, 'discourse in action' (Norris and Jones, 2005). Some, such as those by Jones and his colleagues and by Norris focus on the social processes that lead to the production of creative texts and the social construction of 'creative individuals'. Others, like those of Stockwell and Swann, focus more on processes that go into the interpretation of creative works, Stockwell from a more cognitive perspective and Swann from a more social one. Some, like those of Bhatia and Gillen, deal more with the strategic, socio-pragmatic aspects of discursive action in the context of professional communication and computer-mediated communication respectively. Finally, some, like those by van Leeuwen, Morrison and his colleagues and Jones, invite us to consider the impact of the semiotic resources and technological tools for communication we have at our disposal on our ability to take certain kinds of social actions and engage in certain kinds of social practices.

The third conceptualisation of language in a discourse analytical approach to creativity sees it as part of broader socially informed systems of knowing, being and acting. This conception comes less from linguistics and more from cultural studies and critical sociology, though it has come to occupy an important place in linguistically based methods of discourse analysis. Gee uses the term 'capital D Discourse' to refer to this conceptualisation of language. He defines 'Discourses' as 'ways of being in the world, or forms of life which integrate words, acts, values, beliefs, attitudes, and social identities' (1996: 127). Foucault (1972) and, after him, Fairclough (1992) use the term 'orders of discourse' in much the same way, talking about, for example, the 'discourse of medicine' and 'the discourse of law'.

On the one hand, 'Discourses' or 'orders of discourse' impose constraints on creativity, exerting control over what we can say, what we can think, and the kinds of power relationships that play out in societies. At the same time, 'Discourses' are not fixed. They are vulnerable to being compromised, undermined or transformed as they interact with other 'Discourses'. As Candlin and Maley (1997: 204) note, 'Discourses' consist of 'internally heterogeneous discursive practices whose boundaries are in flux', so as they come into contact with other 'Discourses', 'not only are novel (inter)texts constructed, but novel (inter)discourses arise'.

These transformations occur not only though great works of art or paradigm-changing scientific discoveries, but also through the incremental everyday actions of individuals as they strategically appropriate and combine elements of different 'Discourses' in order to meet the needs of particular moments. Fairclough (1992: 97) argues that 'as producers and interpreters combine discursive conventions, codes and elements in new ways in innovatory discursive events they are cumulatively producing structural changes in the 'orders of discourse'. When discourse is used creatively, it can potentially change 'orders of discourse' on two levels: first on the level of the immediate interaction by shifting the relationships of power among participants, creativity reframing the activity that is taking place, or otherwise creating possibilities for social action that did not exist at the outset of the interaction, and second, on the level of society or culture, by contesting conventional ways of seeing things and opening up possibilities for the imagining of new kinds of social identities and new kinds of social practices (Jones, 2010). Thurlow (this volume) captures the spirit of these small and subversive, though nonetheless profound, acts of creativity in his invocation of Michel de Certeau, who wrote:

> Every culture proliferates along its margins. Irruptions take place that are called 'creations' in relation to stagnancies. Bubbling out of swamps and bogs, a thousand flashes at once scintillate and are extinguished all over the surface of a society. . . . Daily life is scattered with marvels, a froth on the long rhythms of language and history that is as dazzling as that of writers and artists.
>
> (1997: 139–142)

It is chiefly this conceptualisation of discourse that helps a 'discourse and creativity' approach make the connection between 'small c creativity', those tiny everyday creative actions we take with 'small d discourse', and 'big C Creativity', the 'world-changing' aspect of creativity through which new 'big D Discourses' are formed and transformed.

All of the chapters in this volume engage to some extent with this dimension of discourse, considering how texts and the social actions associated with them fit into and interact with broader social formations and systems of value. Stockwell, for example, discusses how creative texts press readers into taking ethical stances and how reading itself becomes a kind of moral act. Similarly, both Stockwell and Swann consider, each from their different perspectives, how engaging with creative texts is not just a matter of resolving meaning but an experience of 'world-building'. One could hardly find a better example of how 'Discourses' are mixed to form creative new '(inter)discourses' than Bhatia's work, reported here and elsewhere (Bhatia, 2008) on the strategic mixing of the discourses of law, accounting, finance and public relations in corporate disclosure reports. Thurlow, in his chapter, provides an excellent illustration of discursive contestation in his description of how the authentic, vernacular creativities of young people using computers are resemiotised by the mainstream media and commercial and educational institutions as 'exotic and outrageous, foolish and pointless, offensive and menacing'. Finally, both Norris and Jones concern themselves with how creative products such as paintings and skateboarding videos function as cultural tools for the formation of individual and group identities spanning timescales from the discrete moment-by-moment actions of everyday life to the longer timescales of 'artistic careers.'

In a sense, it is this engagement with broader issues of social and institutional practices and power, what Thurlow (this volume) calls the 'cultural politics' of creativity, that most distinguishes a discourse approach to creativity from more language-based approaches. It is an approach which, as van Leeuwen (this volume) points out, must be both descriptive and sociological, must endeavour to explain not just 'how people produce and use semiotic resources, but . . . also . . . how these uses come about, how they are taught or otherwise acquired, regulated, debated, (and) changed', and how 'new semiotic resources and practices and new uses of existing semiotic resources are invented'.

Discourse Analysis as Creativity

The contributors to this volume not only illuminate the relationship between discourse and creativity in a wide range of diverse domains from literary reading to jazz improvisation, they also demonstrate the creativity of discourse analysis itself as it has developed over the past half century and continues to develop. In recent years the field of discourse studies has significantly broadened its scope by forging interdisciplinary bridges with sociology, cultural studies, social practice theory, visual communication studies, media studies and cognitive psychology. Consequently, it has taken on board new concerns and priorities, many of which are represented in this volume.

Discourse analysis has, for example, become increasingly interested not just in how texts are put together but also in how people interpret and use discourse in situated social interactions, a theme taken up by nearly all of the authors represented here. It has also become more interested in issues of identity construction, issues which are featured in the chapters by Norris and Jones. It has to some degree also participated in the recent 'cognitive turn' in the social sciences (Stockwell, this volume), evidenced in the chapters by Stockwell and Forceville. It has, in addition, increasingly come to acknowledge the importance of modes other than language in the production of meaning, illustrated by the chapters by Forceville, Norris and van Leeuwen. Finally, it has started to explore what happens to meanings, social practices and social identities when they are mediated through digital technologies, an issue addressed by Thurlow, Gillen, Morrison and his colleagues, and Jones.

In fact, for most of the scholars represented in this volume, discourse analysis as they practise it is itself an example of a creative (inter)discourse, an inventive blending of theories and insights from multiple fields of human inquiry. These new, hybrid approaches to discourse often demand the development of innovative new ways of working which involve mixing text analysis with more ethnographic engagement with people as they go about undertaking their everyday acts of creativity.

What this volume shows is not just that discourse analysis has something valuable to add to our understanding of creative practices and creative processes, but that the study of discourse, indeed the study of language itself is, as Chomsky noted nearly fifty years ago, ultimately and fundamentally the study of creativity. As van Leeuwen points out in his contribution to this volume: 'the semiotician and the artist travel along parallel paths and they might as well talk and work together'.

References

Austin, J. L. (1962) *How to do things with words: The William James Lectures delivered at Harvard University in 1955*, J. O. Urmson (ed.). Oxford: Clarendon.

Bhatia, V. (2008) 'Creativity and accessibility in written professional discourse'. *World Englishes* **27** (3/4) 319–26.

Black, E. (2006) *Pragmatic stylistics*. Edinburgh: Edinburgh University Press.

Boden, M. A. (2004) *The creative mind: Myths and mechanisms* (2nd edn). New York: Routledge.

Bohm, D. (1998) *On Creativity*. London: Routledge.

Candlin, C. N., and Maley, Y. (1997) 'Intertextuality and interdiscursivity in the discourse of alternative dispute resolution' in *The construction of professional discourse*, Gunnarsson, B-L., Linnel, P. and Nordberg, B. (eds). London: Longman, 201–22.

Carter, R. (1999) 'Common language: Corpus, creativity and cognition'. *Language and Literature* **8** (3) 195–216.

Carter, R. (2004) *Language and creativity: The art of common talk.* London: New York: Routledge.

Carter, R., and McCarthy, M. (2004) 'Talking, creating: Interactional language, creativity and context'. *Applied Linguistics* **25** (1) 62–88.

de Certeau, M. (1997) *Culture in the Plural* [trans. T. Conley], Minneapolis: University of Minnesota Press.

Chomsky, Noam (1965) *Aspects of the theory of syntax*. Cambridge, Massachusetts: MIT Press.

Cook, G. (1994) *Discourse and literature: The interplay of form and mind.* Oxford: Oxford University Press.

Cook, G. (1998) 'Linguistics and language teaching' in *The encyclopedic dictionary of applied linguistics*, Johnson, K. and Johnson, H. (eds). Oxford: Blackwell, 198–207.

Cook, G. (2000) *Language play, language learning*. Oxford: Oxford University Press.

Crystal, D. (1998) *Language play*. Harmondsworth: Penguin.

Csikszentmihalyi, M. (1990) *Flow: The psychology of optimal experience.* New York: HarperCollins.

Csikszentmihalyi, M. and Sawyer, R. K. (1995) 'Creative insight: The social dimension of a solitary moment' in *The nature of insight*, Sternberg, R. J. and Davidson, J. E. (eds). Cambridge, MA: MIT Press, 329–63.

Fairclough, N. (1992) *Discourse and social change*. Cambridge: Polity Press.

Foucault, M. (1972) *The archaeology of knowledge & The discourse on language* (A. M. Sheridan-Smith, trans.). New York: Pantheon.

Fowler, R. (1996) *Linguistic criticism* (2nd edn). Oxford: Oxford University Press.

Getzels, J. W. and Csikszentmihalyi, M. (1976) *The creative vision.* New York: Wiley.

Gruber, H. E. and Davis, S. N. (1988) 'Inching our way up Mount Olympus: The evolving-systems approach to creative thinking' in *The nature of creativity*, Sternberg, R. J. (ed.). New York: Cambridge University Press, 243–70.

Harrington, D. M. (1990) 'The ecology of creativity: A psychological perspective' in *Theories of creativity*, Runco, M. A. and Albert, R. S. (eds). Beverly Hills, CA: Sage, 143–169.

Harris, Z. (1952) 'Discourse analysis'. *Language* **28** (1) 1–30.

Jakobson, R. (1960) 'Linguistics and poetics'. Cambridge, MA: MIT Press, 350–77.

John-Steiner, V. (1992) 'Creative lives, creative tensions'. *Creativity Research Journal* **5** (1), 99–108.

Jones, R. H. (2010) 'Creativity and discourse'. *World Englishes* **29** (4) 467–80.

Leech, G. and Short, M. H. (1981) *Style in fiction: A linguistic introduction to English fictional prose*. London: Longman.

Magnusson, L. (1999) *Shakespeare and social dialogue: Dramatic language and Elizabethan letters*. Cambridge: Cambridge University Press.

Maybin, J. and Swann, J. (2006) *The art of English: Everyday creativity*. Basingstoke, Hants: Palgrave Macmillan.

Maybin, J. and Swann, J. (2007) 'Everyday creativity in language: textuality, contextuality and critique'. *Applied Linguistics* **28** (4) 497–517.

Norrick, N. (2000) *Conversational narrative*. Amsterdam: John Benjamins.

Norris, S. (2004) *Analyzing multimodal interaction: A methodological framework*. London: Routledge.

Norris, S. and Jones, R. (eds) (2005) *Discourse in action: Introducing mediated discourse analysis*. London: Routledge.

Pratt, M. L. (1977) *Toward a speech act theory of literary discourse*. Bloomington, Indiana: Indiana University Press.

Runco, M. A. (1990) The Divergent Thinking of Young Children: Implications of the Research. *Gifted Child Today (GCT)*, **13** (4) 37–39.8.

Simonton, D. K. (1988) *Scientific genius: A psychology of science*. New York: Cambridge University Press.

Toolan, M. (1998) *Language in literature: An introduction to stylistics*. London: Hodder Arnold.

Vygotsky, L. S. (1978) *Mind in society: The development of higher psychological processes*. Cambridge, MA: Harvard University Press.

Widdowson, H. G. (1975) *Stylistics and the teaching of literature*. London: Longman.

Woodman, R. W. and Schoenfeldt, L. F. (1990) 'An interactionist model of creative behavior'. *Journal of Creative Behavior* **24** 279–90.

PART ONE

Literary creativity

The three chapters in this section traverse ground that is not unfamiliar to those working in the field of applied linguistics: the linguistic analysis of literary creativity. Each, however, presents a very different route through this territory, and the three different maps they present to us illustrate nicely some of the different orientations towards discourse outlined in the first chapter, Toolan focusing on detecting patterns in literary texts beyond sentence level, Stockwell concerning himself with processes of interpretation as individual readers interact with literary texts, and Swann approaching the reading of literature as a matter of the social practices engaged in by different communities of readers.

Toolan's chapter, coming as it does first in the volume, underlines the debt any discourse analytical approach to creativity must pay to traditional stylistics, which Toolan defines as the study of *patterns* in literary texts. Like the scholars of the 'language and creativity' approach described in the previous chapter, Toolan equates literary creativity with 'literariness', a particular property of texts characterised by the skilful use of various poetic devices, the most important of which for him is repetition. The question he asks is: what makes some forms of repetition more skilful or more 'literary' than others? In the process of answering this question he addresses one of the key issues in the study of discourse and creativity, that of *value,* and presents a convincing argument for regarding 'long established cultural modes and practices such as literature' as sites where we might productively consider what creativity 'crucially consists' of that can be detected *in a text itself.* Such an exploration, he suggests, can provide us with a stronger foundation upon which to begin our consideration of the more everyday or non-traditional modes of creativity considered later in this book.

Stockwell's chapter moves us beyond the creativity that can be found in the text itself to consider the creativity inherent in the reading process, a creativity, he argues, that is central even in the interpretation of texts with strongly defined 'preferred readings' such as the Kipling poem he analyses. In this chapter a more cognitive perspective on creativity is introduced, a perspective which will be taken up again later in this book, particularly in the chapter by Forceville. It is a perspective which has its antecedents not just in the cognitive sciences, but also in the work of literary critics such as Wolfgang Iser and of discourse analysts such as Tuen van Dijk. What Stockwell is concerned with in this chapter, however, is not just the 'workings of the reader's mind' but also, and more crucially, the way readers are placed in what are essentially *ethical* dilemmas regarding whether or not to accept the reading positions that literary texts make available to them. In this regard, Stockwell's 'cognitive poetics' firmly embraces the social and 'political' dimensions of cognition, thus resonating with much of the current work in critical discourse analysis.

The final chapter in this section by Joan Swann widens the circumference of analysis even further beyond the text to explore the social aspects of literary reading. Like Stockwell, Swann insists that literary reading is itself a creative practice – that readers *make* rather than *take* meaning from literary texts. Her interest, however, is less in the cognitive aspects of this process and more in the shared, collaborative and negotiated aspects of it as it takes place within groups of readers. Through her analysis of reading groups of various kinds she argues that literary reading is always to some extent 'locally contingent' and 'essentially collaborative'. In making this argument, she returns us to the issue of *value* raised by Toolan in the first chapter, but approaches this issue from the opposite direction, from the social context in which the text is experienced, debated and judged rather than from the text itself. She also introduces a very different sort of methodology from that seen in the first two chapters, one which quite literally goes beyond the text to include ethnographic observations of actual readers interacting around it. While all three of these scholars can be said to be engaged in the analysis of 'discourse', the 'discourse' that is analysed by Swann, consisting of transcripts of the conversations people have *about* literary texts, is very different from the primary texts analysed by Toolan and Stockwell, and, as Swann demonstrates, requires a rather different set of analytical tools and assumptions about creativity.

CHAPTER 2

Poems: Wonderfully repetitive

Michael Toolan

In this chapter I chiefly try to tease out the connections between literary creativity and that seemingly most uncreative of strategies: repetition. I argue that repetition is central to literariness and literary creativity, but that, clearly, not just any repetition will do: some kinds of repetition work brilliantly, we readers feel (they are 'creative'), others we generally agree are disastrous. How to distinguish the wheat from the chaff? Looking at what many have judged to be a fine poem and a feeble one, I will propose that there is all the difference in the world between informative and enriching repetition (often found in great poems) and redundant, movement-destroying repetition (equally often found in bad ones), and that it is possible to make some progress in specifying the nature of informative repetition and that of what will be judged redundant repetition.

Whatever interim conclusions I reach, it seems to me that we cannot finally exclude contingent judgements of value from the picture, and that such valuation is ultimately a process undertaken by individuals and groups, encouraged to value particular segments of discourse in specific ways by other embedding valuations; immensely dense and complex networks of framing assumptions are at work. But so robust are these framing structures that they can lead (or enable) large constituencies to agree on the high value and high creativity of one poem and the much lower value and creativity of another. I add this to rebut the charge that all evaluative talk of good and bad poetry is either arbitrary discrimination or pure ideological elitism.

Difficulties with 'creativity'

For many, creativity is linked to the changing and breaking of rules, and by extension perhaps the breaking out of normal, ordinary, usual ways of doing things. That is certainly one valuable way of looking at metaphor, one of the core means of creativity in language: fresh metaphor always involves a deautomatisation as the Russian formalists called it, a making-strange of things, events and situations which we have become so familiar with that we no longer really see them; the metaphorised representation makes us see them anew and differently, as if for the first time. There is also the assumption that creativity is a matter of norm- or habit-breaking in Vijay Bhatia's contribution to this volume, where he argues that interdiscursivity is creative, in artfully disturbing accepted patterns of genre and professional practices. Hybridity would appear to be an important kind of creativity, because it is a making of a new form of discourse out of two or more already mastered discourses – one that is noticeably distinct from the master discourses (hence, a hybrid). This is in implied contrast with simple reproduction of a single guiding discourse, 'normal' unhybridised discourse. But there is a danger here of overlooking the potential creativity of the non-hybrid. From Darwin we have learned that, in nature and the creation of species, hybrids (often infertile) are much less important than reproduction with variation, and selective adaptation. Of course there is no intentionality in Darwin's picture of species-creation, so the analogy is inexact, but blending and hybridity are only one form of creativity rather than its essence.

Where does the urge to hybridise come from? In part from technological, social, political and economic changes (these are clearly interdependent) and our cultural and aesthetic adaptations in light of such changes. A simple literary instance of the latter is the emergence of audio fiction, a rapidly growing means of literary consumption, made possible by the massive spread and growth of computers and digitisation, and shrinking comparative costs. For younger MP3 users especially, there is a real choice open to them today as to whether they take in the latest work by Jhumpa Lahiri, Lorrie Moore or Junot Diaz by reading it off the page, or listening to it through their earphones. What adaptive consequences might this bring? How might the short story writer, who knows that their audience will indeed be chiefly an audience and not a readership, judge that they need to adjust their narrative discourse accordingly? There is every likelihood of some creative adaptation in literary practices here.

Charles Forceville (this volume) makes the important point that the assessment of creativity is always *post hoc*, and that we don't seem to be able to provide rules for creativity. As Forceville implies, creativity often seems to be a novel and satisfying solution to a situated problem. Creativity becomes our term for the happy fitness of some new solution to a new or emergent problem, something we had not fully recognised *was* a problem.

The overriding difficulty of the term creativity is that it may have become one of those words or ideas so variously used and understood that it is now of limited value in any analytical enquiry (as distinct from being useful in various rhetorical discourses). We would all prefer to be creative rather than not, we want schools and universities to nurture creativity, and we want creativity in our jobs rather than dull or grinding repetitive servitude. Most of us – perhaps all human beings – also like to be creative in the sense of making things or causing things to happen. You repair something and thereby make it work properly again; or you organise a party in such a way that a wonderfully stimulating atmosphere is created; or you prepare and rehearse a piece of music or theatre, leading to a creative performance; or you make a quilt or grow some vegetables or refurbish your kitchen (a creative makeover). In most of these, the crucial thing seems to be a 'making', the root meaning of the Greek word for poetry; is creativity then just some kind of dependent variable of making? Taking these considerations alongside recent reminders (e.g. in Carter, 2004) that there is plenty of creativity in everyday language and conversation, we might end up deciding that creativity is too general and pervasive, and comes in too many different forms, to be amenable to systematic study or analysis.

To make any progress grappling with *literary* creativity, I believe I need to start from the simplest of first principles. This might involve considering, for example, the seemingly least creative discourse and seeing if it might have an important creative side. But even here, I think, we fail. We have no stable, context-free extra-discursive ground from which we can identify something as radically uncreative discourse. Take silence, for instance: this, we might suppose, must be a good candidate for near-total lack of discourse creativity. For example, you are in a conversation with others, and you say nothing at all: surely the nadir of discursive creativity? Well . . . perhaps not. Silences and pauses, as Beckett and Pinter and John Cage have shown, may speak volumes where words or sounds would have been so much white noise. And who has not used silence, as a communicative strategy, sometimes subtler than words, in their professional or their private life?

A further question will inevitably arise, regardless of what I propose lies at the heart of creativity of the literary kind, namely: what is the relation between whatever is claimed as the essence of literary creativity and whatever is felt to be central to discourse creativity? Is there at least a family-resemblance overlap between the two? As will become clear, this chapter has relatively little to say about discourse analysis directly and makes only limited use of discourse-analytic models and categories; so such questions need to be taken up elsewhere. All I would say here is that in my recent corpus stylistic study of narrative progression (Toolan, 2009) I drew on a constellation of discourse-analytic principles taken from Sinclair, Labov, Hoey, Grice, and others, and that I believe those principles (e.g. collocation and collocability and lexical patterning) are powerful enough to help explain, in different degree, the creativity found in all kinds of discourse.

Creativity, invention, literariness, repetition

A host of stylisticians and linguists with literary interests have reflected on creativity in recent years, including Carter (2004), Pope (2005), Attridge (2004), Hoey (2005), Goodman and O'Halloran (2006), and Dillon (2006). Of particular theoretical interest is Attridge's argument as to why literature is 'singular': literature requires creativity, originality, and invention, but the third of these most of all. For Attridge, creativity is any making of new things (as distinct from making things new), even if the new thing is much like many similar things made by oneself or others previously; and originality is the faculty of being innovative relative to the surrounding culture, by making or doing something which challenges that culture's norms and habitual ways of doing or making; the result is that the product is new not just to the maker – as in creativity – but also to the culture. When we come to invention, however, and the inventiveness of literary art, Attridge argues we encounter not just the ephemeral challenge-to-cultural-norms that originality enables, but a more permanent and enduring challenge: literature does not decay or erode or lose its power over time (a power to move, inspire, even – despite multiple readings – to surprise). In this sense literary texts' significations are *comparatively* free of the controlling and shaping forces of history. They are therefore works which do not permit comprehensive explanation nor require historical analysis as a precondition of understanding. 'To respond to a work of art as inventive,' says Attridge, 'is to share *in the present* something of the inventiveness whereby it was brought into being' (Attridge, 2005: 391).

Attridge's formulation here resonates quite strongly with Margaret Boden's proposal, in *The creative mind* (2004), that the most interesting and far-reaching kind of creativity is 'impossibilist' (rather than only improbabilist) and historical (rather than only personal): 'novel ideas that not only did not happen before but which – we intuitively feel – could not have happened before'. They are 'impossibilist' in that they involve reference to some generative system, but also involve a departure from that system in some way that is not licensed. So such radically creative products are both rule-attentive and rule-breaking. In Boden's terms, they involve the 'mapping, exploration, and transformation of conceptual spaces'. This in turn sounds very familiar to how philosophers and critics for decades (not just since the advent of Fauconnier and Turner) have characterised metaphoricity. My main resistance to this concerns the possible narrowing here of creativity to a chiefly *conceptual* phenomenon. That may fit Boden's interests (her book, after all, is on the creative *mind*), but it may neglect other kinds of embodied, discursive and interactional creativity.

But if from Attridge we take the reminder that literature is especially marked by its singularity of invention, I want to set this now alongside the seeming paradox that literary creativity depends formally on exceptionally adept uses of repetition. Repetition is crucial to literary works in at least two senses. The 'external' sense in which repetition is crucial to them is that, more than any other kind of text, literary texts bear re-reading, performing again, reinterpreting, and intertextual uptake in other works, without wearing out. The 'internal' sense of greater interest here is that the literary text tends to exploit more kinds of repetition than any other kind of text known to me. This is not all it does, of course; for example, a literary text is normally also expected to 'say something', project an idea or a picture, as it were. Even in its invention, however, I will argue that the literary text is oriented to repetition – e.g. to inventively casting a later phrase or structure as echoically linkable to a previous one, thereby achieving (over and over again) a focusing and depth of texture again atypical in non-literary discourse. The achieving of 'depth' logically follows from this saying more about one focused-upon thing (or set of things), rather than shifting the attention to something thoroughly different.

Like any other grand resource (metaphor, analogy), repetition takes many forms, and powerful instances need to be distinguished from failing ones – just as use of metaphor is no guarantee of quality if your metaphors are dead, conventional, mixed, or otherwise misfire. But repetition appears even more vulnerable to failure than metaphor, since we ordinarily associate repetition with being thoroughly uncreative and uninteresting, as can be confirmed by examining the collocates of the words *repetition* and *repetitive*, that is, the verbal company they keep. Googling the phrase *'repetitive and'* (with quotation marks) confirms that *'repetitive and'* is overwhelmingly followed, in the countless texts found on the internet, by such words as *stereotyped, iterative, tedious, pointless* and *monotonous*. Or repetitive is preceded by the verb *become*, with clear contextual implication that becoming repetitive is a bad thing. In all the main phrases within which it collocates, *repetitive* carries a negative semantic prosody. A very similar pattern emerges from the Cobuild Bank of English: after the usual grammatical items (*and, of,* etc.) the highest lexical collocates of *repetitive* are the medically oriented *strain* and *injury,* and then *tasks, work, simple* and *boring.* As for *repetitive and,* there are 124 instances in the 500 million word corpus, noticeably followed by words that include *dull, uncreative, tedious, mechanical, irrelevant, predictable, boring*. Suffice to say that in general terms we regard repetitive activity and repetition in discourse as boring and uncreative, in fact a good indicator of absence of creativity. Someone who produces or enacts a repetition is assumed to be uncreative because, rather than making a new thing, they have made another copy of a previously produced thing.

Over against that negative picture of repetition, I now want to propose that repetition can be immensely creative, and is particularly relevant to literariness and literary creativity. I want to claim that *the right sorts of repetition* almost amount to a design feature or core criterion of literature. My most authoritative source for the idea that repetition is central to literariness – at least in my reading of him – is Jakobson, and his notion of equivalence in his famous 'Linguistics and poetics' article (Jakobson, 1960).

The most powerful and simplest message I take from Jakobson's reflections on language and poetry is that a fundamental principle of verbal art is repetition ('full' or 'partial'). Jakobson himself did not quite nominate repetition as at the defining core of verbal art. He said that verbal art arose where the poetic function was dominant and that one could recognise the poetic function, 'promoting the palpability of signs', by a focus on the message for its own sake. This can be interpreted in various ways, but when it comes to exemplification all Jakobson's instances amount to a valuing of repetition. Why say *horrible Harry* and not *dreadful Harry*, he asks, and then suggests that the speaker without realising it has clung to 'the poetic device of paronomasia,' the latter being a grand rhetorical term for one form of repetition. Similarly the famous *I like Ike* example is essentially built on several types of (partial) repetition. The Jakobsonian dictum specifies as much: 'The poetic function projects the principle of equivalence from the axis of selection into the axis of combination' (1996: 17). If that doesn't mean repetition (albeit creative repetition, repetition with variation), I don't know what it means. A crucial word here is *projects*: units are co-selected in the same syntagma *as if* they were equivalent when they are not, fully (if they were, there would be no 'projection'). The projection of equivalence is a treating of two distinct segments in the verbal sequence as significantly matched, equal or repeated, where the equivalence is partial, not complete. Thus we say that the words *shop* and *top* rhyme, but would not say this of the two words *posh* and *pot*.

So *one* answer to the recurrent questions 'What makes literature literature?' and 'What is fundamental to literary creativity?' is, I believe, that literature exploits and privileges repetition – kinds of repetition, or repetitions with kinds of difference, but repetitions all the same. I think it is not difficult to characterise many literary schemes and tropes, for example, as forms of repetition: rhyme as partial phonic repetition, rhythm and metre as repetition of pulse or beat, assonance and alliteration as consonantal and vocalic repetition, and so on. These are all kinds of repetition within the text; alongside these I would characterise all forms of intertextuality as kinds of repetition, the taking up in a later text of elements used in one or more earlier texts.

One of the main domains of literariness not obviously subsumable under the concept of repetition is that of metaphor and figuration; but even here it may be possible to show that the repetitive treatment of one scene or entity is conducive to the reader's construction of a metaphorical interpretation. For relevance theorists Sperber and Wilson, for example, verbal irony crucially depends on a kind of repetition: an echoic (or interpretively resemblant) mention of an evaluative utterance that is recognised as coming from a prior context and amusingly incongruent in the present context.

That repetition is importantly connected to a text's coherence and whatever theme or macro-proposition a reader might derive in response to the 'aboutness' question was powerfully reaffirmed in Hoey's 1991 book on lexical patterns in text. That brilliant study suggests that those sentences in a text which share the largest numbers of quasi-repetitive lexical links are the most core or central sentences of that text, the ones you would want to keep in an aboutness summary. Hoey there coined the term 'complex repetition', to denote kinds of lexical relatedness which might be via morphological, or potential contextual substitutability, or slightly more distant than that. But bear in mind his focus was exclusively on the lexical. Following Jakobson, I want to include all kinds of repetition, but again, these repetitions are often 'complex', a detectable relatedness (of sound, or rhythm, or word-shape, or meaning) rather than full and simple replication.

Why privilege repetition in this way? Because that in effect is what stylisticians do when they declare that stylistics looks for and identifies *patterns* in the text: a pattern, any pattern, is by definition an instance of some form of repetition. They in turn do so, I believe, because planned, motivated and thoughtful repetition is a *prima facie* simplest demonstration of reflexivity, that is, of text-awareness or of self-awareness. A repeated phrase or line, or a repeated initial consonant on a sequence of words, is extremely unlikely to be produced in 'entirely unmonitored speaking or writing.' Repetition is the simplest and most direct evidence of an elevated reflexivity, or focus on the message (Jakobson), a special attention, over and above what you say, to the way that you say it.

More can be said, however, on what it is about repetition that gives it this special foundational status in literature. Consider the implications of other-speaker repetition in everyday conversation. At a minimum, producing such repetition suggests careful attention to the previous speaker's talk (regardless of whether the repetition accompanies approval or disapproval of that previous speaker or their words). If you can repeat a previous speaker's talk, you have at least demonstrated that you have attended, and quite closely. And that is at the core of what integrational linguists call reflexivity (Toolan, 1996). In this respect at least (but see Dillon's qualifications, below) statistics or frequency are involved, as Firth, Sinclair, and Hoey have all said in their own way, concerning collocation: we call a particular conjunction of words or phrases a collocation, or primed, or a repetition, because the parts have co-occurred in a way that seems far removed from being a matter of mere chance. (In the case of intra-textual repetition, rightly or wrongly, we think of these as absolutely removed from being accidental or a matter of chance; so if we found lines 13 and 14 of a sonnet ending with the words *suit* and *flute*, i.e. coda-repeating rhyme, we would be quite sceptical if the poet confessed she hadn't noticed this, and that it wasn't intentional.)

It is tempting to go one step further and propose that in the canonical speech situation (two people, face-to-face, physically contiguous, with no barriers to hearing or seeing etc.) the only utterance that B *can* produce, in reply to A's initial utterances, so as to leave A reasonably confident that B has recognised and understood what A has said is something close enough to A's utterance to be deemed a repetition. This may be why, in the traditional checking systems on ships and in aeroplanes, the second officer's remarks are extensive repeats of the first officer's. B's being able to repeat A's utterance isn't the whole story, but it is certainly the beginning of a demonstrated understanding. Whereas if B's response to A's instruction is anything other than a full repetition, A cannot be sure that B has attended to all and only what A has said.

As teachers and parents, we know all this; we know that when communicational understanding and compliance get sticky, one useful first strategy is to ask the pupil or child 'What did I just say?' Similarly, if, at the dinner table, I say *This tart is scrumptious* and my companion immediately says *This tart is scrumptious* it is so exceedingly unlikely that the second remark is spontaneous and unaware of my preceding comment as ordinarily to be excluded from consideration. Instead I assume my companion's remark is a designed and calculated repetition, for whatever purpose. There is all the difference in the world between a companion (in the given circumstances) who echoes me by saying *This tart is scrumptious* and a companion who says *This one is stale* or *I don't like mine*. In the first, repetitive scenario, a bond or relation with the prior utterance is assumed, and I embark on some extra inferential work on my remark and her remark: Is there something wrong with what I said? Is *her* food inedible? Or did I miss something else that she just said or signalled? Suddenly an innocent remark becomes part of a much richer text.

Repetition, when looked at in all its productive variety, is at the core of many of the principles and terms which are central to stylistic analytical discourse: marked and unmarked, foregrounded and backgrounded, what is striking or deviant and what is not, what is a pattern and what is unpatterned or seemingly arbitrary. Perceived repetition, identified and nominated repetition – or its equally noticeable opposite, absence of repetition where repetition is expected – underpin all of these.

This argument can be combined with the message that comes down through studies of collocation, from Firth through Sinclair and now Hoey. Rather than imagining that 'rule-bound creativity' gives rise to speakers having creatively to make an open choice at numerous points in every sentence, collocational principles imply plenty of pre-choosing, and the formation of utterances and sentences from prefabricated chunks of wordage, known through countless previous encounters. As Sinclair emphasises, this means we may be making word-selection choices rather less continually than we think. And Hoey, for example, argues that in all sorts of ways we are, thankfully, *not* called upon to be linguistically creative: instead we can rely on extensive expectations, held in memory, about what words, in what combinations, to use in particular recurring and typical circumstances. We are massively primed about language chunks and their discoursal and generic deployment, and in fact to be so primed is what is entailed in being proficient in a language.

At the same time we must beware of reducing literary creativity, in contrast with the foregoing, to novelty or 'unnaturalness' of collocation; in principle creativity might arise at many other levels – and collocational strangeness might not be involved at all. As Dillon (2006) points out, literary creativity can often be achieved by ingenious redeployment of hackneyed and clichéd phrasing within a startlingly unforeseen larger design (the arresting use of clichéd language in altered circumstances is often and powerfully found in the poems of, for instance, Geoffrey Hill, Tony Harrison and Philip Larkin). Dillon warns against assuming – as studies of rare collocations and the semantic deviance in some poetry might tempt us to do (e.g. Hoey on Dickens's use of *ruin(ed)* in *Pictures from Italy*: Hoey, 2005: 173–6) – that the exceptional, inventive or 'striking' phrase is 'proof' of literary creativity. Dillon also warns us against assuming that the presence of clichés in themselves is indicative of bad writing or the non-poetic. Both the rare and the clichéd may figure in literary art, where the key criteria are aesthetic form and design:

> More important than striking, however, is apt, and that is an aesthetic category, not a statistical one and not reducible to one, since it applies to a particular, indeed, a unique communicative moment. And the apt word or phrase may not be a novel one.
>
> (Dillon, 2006: 102)

Using a Wallace Stevens poem as exemplification, Dillon argues that corpus or collocational rarity – exceptionality – may make a line of verse strange but will not guarantee its literariness or that it is art. And he suggests that Stevens's lines achieve literariness not by striking strangeness, but by 'mastering repetition' and by presenting 'the spinning of falling leaves in an eccentric measure that repeats with variation' (Dillon, 2006: 102). I turn at once to considering how a poem by Seamus Heaney masters repetition, while one by William McGonagall fails to do so.

Repetition in 'St Kevin and the Blackbird'

St Kevin and the Blackbird

And then there was **St Kevin** and the **blackbird**.
The **saint** is kneeling, arms stretched out, inside
His cell, but his cell is narrow, so

One turned-up palm is out the window, stiff
As a crossbeam, when a **blackbird** lands
And lays in it and settles down to nest.

Kevin feels the warm eggs, the small breast, the tucked
Neat head and claws and, finding himself linked
Into the network of eternal life,

Is moved to pity: now he must hold his hand
Like a branch out in the sun and rain for weeks
Until the young are hatched and fledged and flown.

*

And since the whole thing's imagined anyhow,
Imagine being **Kevin**. Which is he?
Self-forgetful or in agony all the time

From the neck on out down through his hurting fore-arms?
Are his fingers sleeping? Does he still feel his knees?
Or has the shut-eyed blank of underearth

Crept up through him? Is there distance in his head?
Alone and mirrored clear in love's deep river,
'To labour and not to seek reward,' he prays,

A prayer his body makes entirely
For he has forgotten self, forgotten **bird**
And on the riverbank forgotten the river's name.

from Seamus Heaney, *The Spirit Level*
(Faber, 1996), bold added

How much repetition is there in this poem, and how important is it? We can begin by simply registering some of the poem's most noticeable lexical repetitions. In the version of the poem reprinted above, I have put certain words in bold simply to highlight a preliminary kind of repetition: the lexical repetition links between the title and all subsequent lines. There are a total of eight links, which do not include the indirect repetitions implicit in pronouns. Thus I am not counting here the many places where a *he* or *his* substitutes for a full repetition of the saint's name. But in some ways more interesting are the remarkably local full repetitions (where a more expectable pronominal substitution has not been used: e.g. lines 2–3, which include 'inside/ His cell, but his cell is narrow'). And then there is the repeating with difference in one stanza what has already been reported in the previous one: as when stanza 1 describes the saint's *arms stretched out* and stanza 2 partly tells us this again: *One turned-up palm is out the window*. This in turn is reformulated in stanza 4: *now he must hold his hand/ Like a branch out in the sun and rain*. When you are told things this often, you begin to believe they are true; until you are told the whole thing's imagined.

Another kind of repetition are the triples, of verbs (and their surrounding phrases), with concomitant ellipted elements (cf. Dowis, 1999: 117, who claims 'the human ear has a peculiar affinity for triplets' and the insight attributed to Martin Joos, that it takes three to make a row). That is to say, in some of these it is the ellipted material which is the repeated material (as in the first example below, where *a blackbird* must be retrieved as subject of *lays* and *settles down*), while in others some ellipted repeated material occurs alongside some overt repetition (as in the final example below, with ellipted *he has* and overt *forgotten*; this example also shows how one kind of repetition does not preclude another, since it contains repetition of *river* in addition):

a blackbird **lands**
And **lays** in it and **settles down** to nest

Until the young are **hatched** and **fledged** and **flown.**

he has **forgotten** self, **forgotten** bird
And on the riverbank **forgotten** the river's name.

There are many half-rhymes and assonantial repetitions, subtly arranged in the poem; I will point to just a few. While there is no orthodox line-rhyming here, it feels motivated rather than accidental that the first and last tercets of the first half of the poem end with *so* and *flown*, while those of the second half end with *time* and *name*. And more localised forms of sound recurrence are surely at work in the phrase *linked/ Into the network of eternal life*; or, a line earlier, in the assonance of *warm eggs* and *small breast* (these vowels echoed in reverse order towards the end of the third thing Kevin feels here: *the tucked/Neat head and claws*).

Set against these and other repetitions in the poem, there are elements that are striking and foregrounded, or distracting attractors, by virtue of *not* being a repetition or an echo or predicted by anything that has gone before. One such, in this short poem of straightforward vocabulary, is the nonce word *underearth* (not to be found with this sense, and unhyphenated, in the OED). *Underearth* is so much not a repetition that it arguably does not even exist as a recognised word in English; and this unrepeating originality is compounded by the phrase, full of connotations but quite opaque semantically, within which it sits: *the shut-eyed blank of underearth.*

Nevertheless, perhaps the most prominent novel element is the poem's final focus, on the riverbank by the river whose name Kevin has forgotten. *Is* Kevin's cell by a river? There is no mention of one in the first four stanzas' description. The only previous mention of a river in the poem was four lines earlier: *love's deep river.* So the river is no concrete entity in the context or world in which Kevin and the blackbird and the cell are concrete entities; the river is no more graspable than the abstract idea, love (or, on a less persuasive interpretation of *love's deep river*, the river is some quality possessed by love). Suffice it to say that, one way or another, in its abstraction, metaphoricality and deictic unanchoredness, you cannot grasp and hold this love-river the way Kevin can have and hold the blackbird and her eggs. So there is a powerful and unrepeatable world-shift when we get to the final line, where Kevin is depicted *as if* kneeling by a real riverbank, blanking on a real river's name. But by this point we are caught up rather than disturbed by the situation, our mental labours their own reward; aided by the poem, we have imagined the whole thing.

At this point in my review of repetition and literary invention, I would like to support another of Attridge's theses about literature: he suggests it enables or compels an enriched apprehension of the Other, and therefore of the Self. That apprehension of the Other comes to Heaney's Saint Kevin, I think, when we are told he is *Alone and mirrored clear,* near the end of the poem. Yes he is alone in a sense; but in his circumstances, this long spot of time he must devotedly maintain, having put all the blackbird's eggs in the one basket of his hand, he is exceptionally mirrored too, and able to see himself especially clearly and detachedly. This is part of the looking and looking again, at the same thing, that creative literary repetition enacts. Nothing in our everyday world more succinctly unites these ideas of repetition and otherness in the visual medium than the mirror (the aural medium counterpart is the wall or cliff face that returns to us an echo). A mirror need not be made of glass with a silvered backing, of course; any plane of relatively still water will do – such as a pond or a river. Like verbal repetition, visual mirroring entails recurrence with subtle difference or otherness (a mirrored face is not quite the face that others, viewing the face directly, see). Even without that complication, in seeing one's face or body in a mirror one sees oneself as an Other, as subtly or shockingly different from what one normally sees oneself as, in a mirror-free world. We need poems as we need mirrors, even if we might conceivably get by without them. The mirror enables an early form of self-reflexivity, that defining property for integrational linguists of humans as language-using and language-making animals.

Now, by contrast, consider the following poem by the nineteenth-century Scottish poet William McGonagall, often described as Britain's greatest bad poet. I reproduce just the first four stanzas of this 14-stanza poem:

A Tale of Elsinore

A little child stood thinking, sorrowfully and ill at ease,
In a forest beneath the branches of the tall pine trees –
And his big brown eyes with tears seemed dim,
While one soft arm rested on a huge dog close by him.

And only four summers had passed o'er his baby head,
And, poor little child, his twin brother was dead,
Who had died but a few days before,
And now he must play alone, for he'd see him no more.

And for many generations 'tis said for a truth
That the eldest bairn of the Cronberg family died early in youth,
Owing to a curse that pursued them for many a day,
Because the Cronberg chief had carried a lovely maiden away,

> That belonged, 'tis said, to the bold Viking chief,
> And her aged mother could find no relief;
> And she cursed the Cronberg family in accents wild,
> For the loss of her darling, beautiful child.

Imagine young McGonagall is sitting in your creative writing class, and this is his first draft submission. It is an interesting exercise trying to decide what you would have him do, and what you would have him do first, to 'enhance the literary value or creativity'. Perhaps remove all the adjectives? And also most of the adverbs? These opening stanzas, indeed the whole poem, display various kinds of lexical and structural repetition; but they differ sharply from those found in the Heaney poem in their effect, as I will attempt to demonstrate with the very opening line of the poem.

What is wrong with *A little child stood thinking, sorrowfully and ill at ease*?

If someone is a child, our default assumption is that they are likely to be *little*. Only if a child is *big*, or *angry* or *tiny* or *overgrown* is the latter information unpredictable enough to be tellable. The *little* in *little child* is like a repetition – but much worse than if a true (and often paradoxically informative) repetition were used. *Little* is a kind of repeating or saying again and redundantly, what is semantically implicit in *child*. The same general objection can be made to *ill at ease*, relative to the *thinking, sorrowfully* that immediately precedes it: if someone is thinking sorrowfully, it more or less goes without saying that they are *ill at ease*. What does the information *ill at ease* add to the picture we have already formed, of the child standing thinking sorrowfully? Arguably nothing: hence, again, a kind of redundancy, restating what was already partly explicit and partly implicit, i.e. again the 'wrong sort of repetition'. The syntagmatic choice is arguably worse than this, if you feel that *ill at ease* is a semantic 'downtoning' from *sorrowfully* (cf. the incongruity of saying *Jenny was very tall and above the average height*).

Similar kinds of redundant repetition or stating of the co-textually obvious occur in line 2, where a stripping away of quasi-repetitive redundancy might begin with removal of the final word *trees*:

> In a forest beneath the branches of the tall pines OR
> In a forest beneath the branches of tall pines

Then the reader wonders whether *In a forest* is needed, whether it significantly informs or enriches the texture beyond what we can understand from:

> Beneath the branches of the tall pines

And then, those *tall pines*: like 'little children', our powerful default assumption is that trees, especially pines, especially pines close by a child standing, will be *tall*! Again, the tallness goes without saying, so that line 2 is further reducible to:

> Beneath the branches of the pines

It may be objected that McGonagall 'needed' *tall* and *trees* and *In a forest* to make the line scan and rhyme and be rhythmical, and to create a fuller more vivid picture. My point is that the first reason cannot *drive* the selection of wordings (the rhymes and rhythms must feel as if they naturally fit, i.e. that the wordings are lexically and thematically or narratorially justified, before we notice that they also work in relation to rhyme and metre) and that the second reason simply isn't met: McGonagall's inclusion of these words is largely redundant – the 'wrong sort of repetition' – and does not enrich or enlarge the picture. Perhaps these opening lines are extreme, but instances like these are found throughout the poem. Typically they are quasi-repetitions (not full ones, of the *Break, break, break* or *Tomorrow, and tomorrow, and tomorrow* variety, see below) but immediate (adjacent) ones; ironically or not, it is just this combination that seems least defensible or acceptable in poetry. The repetitions in the Heaney poem, by contrast, are full if immediate (*his cell . . . his cell; forgotten . . . forgotten . . . forgotten*), or mostly quasi-repetitive if longer distance.

Where there *is* longer-distance full repetition in the McGonagall poem, it is more confusing than enriching, as can be seen in the case of *child*, used three times already in these first four stanzas (lines 1, 6, and 16): the difficulty is that while the *child* of lines 1 and 6 is co-referential, that of line 16 evidently denotes a quite different female individual (and, more confusing yet, a sufficiently older child to attract the attention of the Cronberg chief). My main point is that the repetition in 'A Tale of Elsinore' is literarily faulty, in that it reformulates without yielding new information or contextual effects. Elsewhere, his expressions misfire because they are at odds with natural primed collocational English; or because they add speaker-evaluation where none is needed, the evaluation provided (*which was a terrible shame*) being the unmarked one that any normal reader could produce – so the reader's intelligence is insulted and their scope for engagement and response is excessively restricted.

I have tried to identify some of the ways in which McGonagall's repetitions are the wrong sort of repetition. Among other things, they seem to fail by virtue of being co-textually immediate (adjacent) but not full (i.e. lexically or formally). Immediate repetition, it seems, usually needs to be full, as in these famous examples from Tennyson and Shakespeare:

> Break, break, break,
> On thy cold gray stones, O Sea!

> Tomorrow, and tomorrow, and tomorrow,
> Creeps in this petty pace from day to day
> To the last syllable of recorded time

Full immediate repetition is often entirely acceptable, since it is then an entirely conscious and calculated repetition, a knowing that you are emphasising the already said. The appearance, adjacently, of the same form a second or third time – where utterer's absent-minded error can be discounted – licenses the implicature that some other information or relevance motivates the use, other than the interpretation adequately communicated by the previous initial use.

While such full adjacent lexical repetitions comprise a large set of instances, most lexico-semantic repetition and near-repetition in successful poems is longer distance: a mention of *earth* or *stain* or *flower* at one point in the text is echoed by re-use of the word or an associate (a complex repetition) elsewhere. But in McGonagall, unfortunately, it is not clear that we have much long-distance repetition, and all too much local, redundant-type repetition. The principle seems to be: immediate repetition must be full, not a reformulation, lest the reformulation be deemed ill-judged redundancy. This principle McGonagall repeatedly violates. For example, in line 2 of his infamous poem on 'The Tay Bridge Disaster', McGonagall writes *Alas!* and continues *I am very sorry to say*. This is an example of immediate non-full repetition. It is therefore, interestingly, much worse than, for example, writing *Alas! Alas!* Mishandling of repetition is not McGonagall's only weakness, to be sure. In addition, he has no ear for usage or collocation. In the final couplet of the Tay Bridge poem he declares:

> For the stronger we our houses do build,
> The less chance we have of being killed.

Since *chance* usually co-occurs with good things (we normally speak of the *chance* – not the *danger* – *of winning, passing, succeeding,* etc.), a word like *danger* or *risk* would have been much more appropriate here. Again he has chained words together – *chance, being killed* – in ways that are not collocationally sanctioned, are not primed in the language, and therefore are jarring or unnatural (Hoey's term) in the way that learners' English can sometimes be unnatural (producing words that are grammatical in item-and-arrangement terms, but that lack attention to the idiom principle, or semantic prosody, which is also often necessary – and never more necessary than in literary production).

Conclusion

Creativity, and its more specific forms (literary creativity, musical creativity, creativity in cooking, dance, chess, etc.) may entail inventiveness, ingenuity, originality, adaptation, blending, change, novelty, but these are not sufficient. Similarly, science and technology, wars and finance continually change our environment and us, so that with the emergence of hand-held computers, i-pads and video-phones, many new activity types or forms of life, to which distinct social groups can become enormously committed, can emerge. And academics have not been slow to analyse and theorise these. But composing a collaborative narrative on the internet rather than *viva voce* around a table does not guarantee creativity, any more than deploying the distinct affordances of the electric guitar (some of which are not available to an acoustic one) does; if it is to mean anything, creativity must be more than proficiency. Again, I would agree that innovation and interdiscursivity in dress, speech and behaviour – like their opposites, conservatism and homogeneity – can reform and perform social or personal identity. But it can only be one sense of creativity that is involved in identity-projection; and to the extent that the concept of identity is predicated on a notion of 'sameness' among several members of a population, it must also entail some restraint on degrees of creative difference.

For these reasons I think there is still something to be said for thinking about creativity in relation to a long-established cultural mode and practice, such as literature. For the important fact remains that, even with all the cyberfication encouraged by digital media, people still sit down to write and read poems about someone they have loved and lost, or the way a river runs through a locality and all the lives it contains, or about a new grandchild brought back to the old family home. How can we bear to write and read of these old, old themes and preoccupations, unless it is because something new and different is done in these new offerings, which we summarise as creative? And in what does that creativity crucially consist? That is the question I have tried to begin to answer; my answer has entailed equating literary creativity with literariness and linking both of these with some kinds of lexical repetition, not others. Literary creativity I believe is characterised by comparative increase of internal repetition (central to the work's literariness, and reflected in our assessment of a work's coherence) which nevertheless eventuates in readers' judgements that taken as a whole, in context (the external perspective), the text is to a significant degree *not* a repetition of previous works, but a new and unforeseen departure, a productive variation.

References

Attridge, D. (2004) *The singularity of literature*. London: Routledge.

Attridge, D. (2005) 'A response to Rob Pope'. *Language and Literature* **14** (4) 390–2.

Boden, M. (2004) *The creative mind*, 2nd edn. London: Weidenfeld and Nicolson.

Carter, R. (2004) *Language and creativity: The art of common talk*. London: Routledge.

Dillon, G. L. (2006) 'Corpus, creativity, cliché: Where statistics meet aesthetics'. *Journal of Literary Semantics* **35** (2) 97–104.

Dowis, R. (1999) *The lost art of the great speech*. New York: Amacom.

Goodman, S. and O'Halloran, K. (2006) *The art of English: Literary creativity*. London: Palgrave.

Hoey, M. (1991) *Patterns of lexis in text*. Oxford: Oxford University Press.

Hoey, M. (2005) *Lexical priming: A new Theory of words and language*. London and New York: Routledge.

Jakobson, R. (1960) 'Closing statement: Linguistics and poetics' in *Style in language*, Sebeok, T. A. (ed.). Cambridge MA: MIT Press, 350–77. Reprinted in *The stylistics reader*, Weber, J. J. (ed.). London: Arnold, 1996, 10–35.

Pope, R. (2005) *Creativity: Theory, History, Practice*. London: Routledge.

Toolan, M. (1996) *Total speech: An integrational linguistic approach to language*. Durham, NC: Duke University Press.

Toolan, M. (2009) *Narrative progression in the short story: A corpus stylistic approach*. New York and Amsterdam: John Benjamin.

CHAPTER 3

Ethics and imagination in literary reading

Peter Stockwell

Lying between Egham and Windsor on the river Thames west of London, the water meadow at Runnymede was the site of the sealing of the Magna Carta by King John in 1215. This document is regarded as one of the founding texts of North American, European and British Commonwealth democracy and the rule of law. It is widely cited as enshrining the principles that no authority (including the king) was above due process, that imprisonment without charge or reason was unlawful, that rights and obligations should be written, fixed and binding. Written in the dead and thus unchanging church language of Latin (rather than the French dialects of the king and barons, and certainly not in demotic English), the Magna Carta also implicitly established the notion that a legal document should be precise and unambiguous enough to allow no creative interpretation, readerly variation or misconstrued arguments. It is regarded by many, especially in the United States, as being a direct precursor to the American Constitution and Bill of Rights. A memorial at Runnymede, preserved by the English National Trust, reads:

> This oak tree, planted with soil from Jamestown, Virginia, the first permanent English settlement in the New World, commemorates the bicentenary of the Constitution of the United States of America.
>
> It stands in acknowledgement that the ideals of liberty and justice embodied in the constitution trace their lineage through institutions of English law to the Magna Carta sealed at Runnymede on June 15th, 1215.

The site at Runnymede also hosts a memorial to President John F. Kennedy, a memorial to the Allied air forces of the Second World War, and another oak tree planted by the prime minister of India.

The notion of written and authoritative law that Magna Carta symbolises views any creative interpretation of legal text as a consequence to be diminished and avoided. In this chapter, I examine how the interpretative creativity of the reader can be exploited and manipulated by a certain type of text to produce an ethical interpretative position with which the reader might not necessarily agree nor find satisfaction. I distinguish between the process of reading and its cognitive mechanics on the one hand and the felt effects of an interpretation on the other. I hope to show how even highly defined texts with a strongly preferred delineated interpretation require some creativity on the part of the reader, and also how such literary texts can press a reader into adopting a particular ethical position.

The Magna Carta was both renewed and suppressed several times throughout the medieval and Renaissance period, was held as a supporting document by both sides in the English civil war, and was regarded as the guarantor of parliamentary sovereignty throughout the nineteenth century even as that same parliament repealed most of its provisions from English law. Nevertheless, it remains in general popular perception as a founding and uncompromised statement of rights and liberties. In 1911, the document was celebrated by this now famous poem by Rudyard Kipling.

The Reeds of Runnymede

(Magna Charta, June 15, 1215)

At Runnymede, at Runnymede,
  What say the reeds at Runnymede?
The lissom reeds that give and take,
That bend so far, but never break.
They keep the sleepy Thames awake
  With tales of John at Runnymede.

At Runnymede, at Runnymede,
  Oh, hear the reeds at Runnymede: –
'You mustn't sell, delay, deny,
A freeman's right or liberty.
It wakes the stubborn Englishry,
  We saw 'em roused at Runnymede!

'When through our ranks the Barons came,
With little thought of praise or blame,
But resolute to play the game,
  They lumbered up to Runnymede;

And there they launched in solid line
The first attack on Right Divine –
The curt, uncompromising "Sign!"
    That settled John at Runnymede.

'At Runnymede, at Runnymede,
Your rights were won at Runnymede!
No freeman shall be fined or bound,
    Or dispossessed of freehold ground,
Except by lawful judgment found
And passed upon him by his peers.
Forget not, after all these years,
    The Charter signed at Runnymede.'

And still when Mob or Monarch lays
Too rude a hand on English ways,
The whisper wakes, the shudder plays,
    Across the reeds at Runnymede.
And Thames, that knows the moods of kings,
And crowds and priests and suchlike things,
Rolls deep and dreadful as he brings
    Their warning down from Runnymede!

The poem is widely cited approvingly by conservative, right-wing, and libertarian politicians and observers especially in the USA (see, for example, http://www.freerepublic.com, the blogs callingengland.blogspot.com, http://vanishingamerican.blogspot.com, and http://herbertson.wordpress.com). The last two stanzas, in particular, are widely quoted: they feature twice, for example, in speeches made by the right-wing former British prime minister Margaret Thatcher (on 26 November 1984 in the second Carlton Lecture 'Why democracy will last', and on 11 January 1996 in the Keith Joseph Memorial Lecture 'Liberty and limited government'). Appearing first in a schoolbook history of England (Fletcher and Kipling, 1911), the poem was well known to generations of British schoolchildren who could declaim it aloud from memory; in more recent years it has become less widely used in schools. By contrast with its political and educational prominence, it is surprisingly uncited in scholarly academic work, and any close textual analysis is largely restricted to noticing the shift in verse-form between the first two and the final three stanzas.

In this chapter, I hope to remedy this failing by presenting an account of the poem consistent with *cognitive poetics* – the understanding of literary reading in the light of current thinking about the mind (see Stockwell, 2002, 2009a). This approach is part of a general cognitive turn in arts and humanities research over the last few decades, which has encompassed linguistics and cultural studies. In the pioneering work of Lakoff (1987, 2002) and van Dijk (1980, 2008), for example, cognitive approaches to language can be regarded as a form of discourse analysis with a strong stylistic tradition, though there is also a claim to a psychological and empirical validity. Since its object of study is neither text alone nor mind alone, but an emergent blend of both, cognitive poetics represents a promising approach to the similarly heteronomous phenomenon of creative activity in literary reading.

The poem, then, comes with a rich provenance and set of contexts, and it has a variety of uses in the world. Like the Magna Carta text on which it draws allusively, it appears at first to be highly assertive and fixed, but it is easy to observe a range of readings and uses of the text. Umberto Eco (1972, 1979) describes relatively 'open' and 'closed' literary texts, in the sense that some works allow for a greater readerly manoeuvring of interpretation than others, and the Kipling poem in this regard looks 'closed': it is declamatory and dogmatic, positioned very firmly in its ideology and the perspective of its values. Eco's description of 'closed' texts is a relative category rather than an absolute one. He regards openness as an essential aesthetic property of literariness, which is present even in relatively 'closed' texts such as his analyses of comic books and other low-art examples, and he understands openness as a feature designed and placed into the literary work deliberately and intentionally by the author.

There is no question that the Kipling poem is generally regarded – by both literary scholars and popularly – as a politically positioned and ideologically committed work (see Sullivan, 1993; Gilmour, 2003; Mallett, 2003; and the articles in Bloom, 2004). 'The Reeds of Runnymede' appears in the original collection a few pages before the next poem, 'My Father's Chair', which begins:

> There are four good legs to my Father's Chair –
> Priest and People and Lords and Crown.
> I sits on all of 'em fair and square,
> And that is the reason it don't break down.

Here, the authorial voice usually designated as Kipling expresses a similar ideological reliance on a constitutional balance between church, state, people and monarchy as in the other poems in the 1911 book. Kipling as a biographised literary life is characterised as a populist defender and apologist for jingoistic British imperialism, 'the voice of the hooligan' in the phrase used by his contemporary, writer Robert Buchanan (1899), and the most prominent champion of nationalist Britishness (see Mallett, 2003). 'The Reeds of Runnymede' certainly seems to speak uncompromisingly from an ideological perspective that stands for 'the people' – here including aristocratic barons, rendered on the side of ordinary free men by contrast with the king. Furthermore, this ideological position is presented as unchanging and unstinting ('stubborn Englishry' who are otherwise content and asleep in the poem), and as part of an order of Englishness that is natural: there are 'English ways', and the voice of the people is naturalised as the voices of the reeds on the riverbank.

I have chosen this text for examination in this chapter precisely because its status, meaning and values seem to be so precisely fixed and discernible. There are no developed subversive readings of the poem that I could find in which it is presented as an ironic critique of nationalism, or a sarcastic attack on liberty as a delusion, or as a pro-authoritarian and anti-libertarian tract. I think these readings if they existed would largely be regarded as odd and wilfully and mischievously eccentric, as being 'wrong' in some fundamentally convincing sense. There are, of course, readings that are resistant to its message, but these are largely simply dismissive of the poem's message on poetry-reading blogs and websites, and all of them accept by default the fact that there is *a message* in the poem that is definite, assertive and fixed. In other words, resistance focuses on disputing the poem's underlying ideology, rather than trying to argue for a re-interpretation of the poem that claims it is actually saying something different. There seems little room for interpretative variation or readerly creativity in it, and indeed there is little evidence in the world that any such attempts have been published or disseminated. In this chapter, I would like to show the great extent to which a reader must work creatively even with such apparently inflexible material.

In the terms established in Stockwell (2009a: 161–6), there is a very strong and precisely delineated sense of a *positioned reader* in 'The Reeds of Runnymede'. As in other Kipling poems (see my analysis of 'If' in Stockwell, 2005), a preferred readerly perspective is very assertively constructed by the text. The main mechanism for this is an assumption of perspective and agreed values that is inescapably encoded in lexical and grammatical choices throughout the poem. The style is direct with a second-person address, first as a forceful imperative ('You mustn't sell, delay, deny, A freeman's right or liberty'), and then as categorical assertion ('Your rights were won at Runnymede! No freeman shall be fined or bound, Or dispossessed of freehold ground, Except by lawful judgment found And passed upon him by his peers'). There is a slippage here, of course, from the first 'You' which is both a generalised 'one' as well as a direct address to the king or any other abusive authority, to the second 'Your' which is also generalised but this time to all free men, including you the reader. The desirability of the events is established through lexical choice: 'rights' that were 'won'. Other strongly polarised evaluative markers include 'resolute', 'stubborn', 'solid', and 'uncompromising'. This evaluative lexis not only invites agreement but constructs it in a strongly preferred interpretation that is difficult to evade in the process of reading.

The poem, then, has a strongly delineated ethical frame within which the ideological positions of author and reader are assembled, with a high degree of identification between these two roles during the reading of the poem. (I am using a view of personality here – which includes a sense of ideological self – as flexible, plastic and provisionally 'soft-assembled' in reading: see Gibbs (2003) for a cognitive psychological view, Bucholtz and Hall (2005) for a social constructivist version, and Stockwell (2010) for a cognitive sociolinguistic synthesis between these two). The reader who is positioned while reading 'The Reeds of Runnymede', of course, does not have to buy into a comprehensive identification with the Kipling of his other poems, nor with jingoism or imperialism in general, but if a particular reader's knowledge of 'Kipling' includes all of these aspects, then such a reader will find it difficult to disentangle these different facets. This is what generally happens when historically knowledgeable academic readers encounter Kipling's texts, and their sense of discomfort arises often because of a resistance to the type of reader within which the text is trying to position them. (An account for the difference between academic and non-scholarly readings of Kipling appears in Stockwell, 2009a: 144–52).

An 'open' literary work (in Eco's 1972 terms) allows for a range of readerly positioning, or delineates its readerly construction only weakly. This falls short, of course, of saying that a text can have no imposition of a reading position at all: all literary works, even highly ambivalent, ambiguous, non-dogmatic and playful texts, offer a matrix within which a plausible set of readerly positions can be taken. Given a singular literary work, no such matrix is infinite, though the possible positions it might contain could be very large indeed. By contrast, at the other end of the scale, a relatively 'closed' work like 'The Reeds at Runnymede' offers a narrow range or a very strongly positioned reader-assembly. Whatever the longstanding disposition of the particular reader in terms of knowledge, temperament, ideology or preference, the Kipling poem seems only to offer either a close conformity or an outright rejection (a reading that 'boxes up' the preferred interpretation and disagrees with it as an object). A discomforted sense in a reader comes not from a failure of the text to impose an agreeing readerly position, but from the ability of such a person to be conscious that the constructed positioned reader-assembly does not match their own usual longstanding self-awareness.

The French Marxist literary theorist Pierre Macherey (1978, 1985) has argued that all texts contain 'silences' in the form of what their authors or the conditions of production have determined they should not say. With particular reference to literary works, these silences take several manifestations, deriving from the fact that literary texts are read allusively and intertextually in the context of other literary artifacts. What Attridge (2004) later called the 'singularity' of literature entails that every literary work – even highly formulaic or generic texts – are peculiar to themselves and are therefore both defined and take their value from all the other literary texts that they are not. Macherey's view of the consequences of this is devastating for the institution of literature: he regards its general and essential condition as being consensual and communicative, and thus a fundamentally bourgeois phenomenon inimical to a proper understanding of the materialism of the class-struggle and social conflict (see also Balibar and Macherey, 1978). Aside from the pathological paranoia of this last point, Macherey's notion of silences as a key property of literariness seems evidential.

> To explain the work is to show that, contrary to all appearances, it does not exist for itself, but, on the contrary, bears within itself the mark of a determinate absence which is also the principle of its identity: made hollow by the allusive presence of those books in the face of which it is constituted, turning upon the absence of what it cannot say, haunted by the absence of certain words to which it returns again and again, the book is not built upon the development of one meaning, but out of the incompatibility of several meanings, which is also the bond which relates it to reality in a prolonged and ever renewed stance of confrontation.
>
> To see how a book is made is also to see of what it is made: and what is it made of if not that lack (*défaut* – break, difference, flaw) which gives it its history and its relation to history?
>
> (Macherey, 1978: 98, translation note from cited edition)

The argument prefigures Wolfgang Iser's (1974, 1978) similar but reception-oriented notion that literary texts consist largely of 'indeterminacies' and 'gaps'. The literary text itself is not sufficient to account for the richness of its effects in the reader, and it is in the inferencing, connecting, enriching and fulfilling of the text in the reading process that the literary text becomes the literary work (Iser develops these notions from his former teacher

Gadamer (1989) and the philosophy of Ingarden (1973a, 1973b)). In his later thought, Iser (2000) describes the interpretative act as a 'figuration' between text and reader, in which reading is a performance (of identity, as also suggested above by a cognitive poetic understanding of the plasticity of the readerly role). The literary work arises not as a property of textuality but as an 'emergence' in the reading process, and literary reading becomes not simply the resolution of meaning but the experience of world-building.

In spite of the plausibility in general of these ideas, the critical theory never gets around to describing the psychological mechanics nor the stylistic configurations that converge to produce literary work out of literary text. It is immediately clear that a complex and ambivalent literary text always massively underdetermines most readings of it, which must therefore be derived from readerly input of various types. It is less clear how a relatively closed text such as 'The Reeds of Runnymede' provides sufficient indeterminate space for a reader to engage creatively. Given the passionate and intensely held opinions about the poem in non-scholarly arenas, there is no question as a simple empirical matter that the poem generates powerful and resonant responses. How can these be accounted for systematically and in accordance with the principled insights of cognitive poetics?

The first thing to observe about 'The Reeds of Runnymede' is that there are internal, grammaticalised 'gaps' (opportunities or requirements for readers to engage in some local inferencing work) and also external, historical 'silences' (which we would usually regard as 'context', though here I include readerly dispositions of emotion and temperament as well as matters of knowledge and literary history). In cognitive science, it has been established that, for example in the visual field as in all perception, what we see is far richer than the raw stimulus information that hits our eyes. Our rich cognitive processes go to work on the raw data almost immediately, inverting the lensed image, filling in colours, shapes and functions from memory and guesswork at the edges of our vision, assigning values and object-permanence to gestalt forms, keeping track of variable shapes, and predicting the likely configuration of the next input so as to be more efficiently ready for it. We differentiate perceptual objects not wholly on the basis of their objective properties but significantly also on the basis of our prior experience of such objects, so that what we see, focus on and maintain interest in is considerably conditioned by cognition and emotion.

It is a principle of cognitive science that there are continuities between our different perceptual and experiential systems, so the way we discern figure and ground in the visual field is essentially the same process for the auditory, tactile and other sensual fields; equally, higher level conceptual experiences (such as reading a literary text) draw on exactly the same capacities and processes. Grammaticalised and textualised 'gaps' in a poem, then, can be understood in a similar way to our current best understanding of how non-figure absences are perceived in the field of attention.

Useful recent work on attention has been undertaken by Smith *et al.* (2003), Peterson and Enns (2005) and Carstensen (2007), adapted for literary reading by Stockwell (2009a, 2009b). Figures in visual perception are usually regarded as positive 'blobs' or shapes, with a greater or lesser degree of granularity according to the depth of focus and intensity of attention. These objects have edges, and the edge is processed along with the shape: in other words, the edge of a figure belongs to the figure, rather than to the ground that surrounds it. However, there are also cases in which the figure has been removed from the ground or is otherwise understood to be absent: holes, dents, the space between objects that is the focus of attention, and figures that disappear or are destroyed are regarded as 'negative blobs' (Carstensen, 2007). Such negative figures *only* have edges. A negative figure can be understood as a *lacuna* when it occurs in literary reading, and the effect of such lacunae is often one of readerly discomfort and a sense of 'felt absence' (Stockwell, 2003, 2009b). The sensation is of something unspecified or removed, rather than never having been mentioned or evoked at all. Lacunal effects in reading produce feelings such as uncertainty, vagueness, unspoken threat, or delicate and rarefied perception that is just at the bare level of consciousness: in short, the resonant sensations of gaps and indeterminacies that feature in all literary discourse.

Lacunae are the immediate textural effect of negations, most obviously, and there are many negational forms in 'The Reeds of Runnymede', perhaps surprisingly for such an assertive and positive poem. Aside from the grammatical negations (about which more later), there are examples of qualitatively negative lexis ('take', 'stubborn', 'blame', 'bend', 'rude' and even 'sleepy'), where the negative aspect of these is pointed up often contrastively in a pair ('give and take', 'praise or blame', 'sleepy Thames awake'). There are also lexical negations, in the forms 'little thought', 'never break', 'No freeman', and morphological negation in 'uncompromising' and 'dispossessed'. The weight of these negations is in the first half of the poem, and they are replaced contrastively by the strong positive assertions in the middle stanza, where the only negative element is 'uncompromising'. The lacuna effect of this early cluster of negations is to raise the resonant memory of broken reeds (metaphorically people), the sleepy Thames (metaphorically the country), of rights and liberties sold, delayed, denied in previous times, and of the compromise that was a possibility that was not realised.

The different types of grammatical negation are also significantly arranged. The first, 'never break', conveys an emphatic positive sense through a combination of grammatical and semantic negation (a lacuna with strongly defined edges). The second negation ('You mustn't sell, delay, deny . . .') sets out, within direct speech, the prohibitions of the charter, but it is the obligation (deontic) modality of 'must' that is more prominent than the reduced negative particle 'n't'. The uncompromising assertiveness of the middle stanza contains no grammatical negation at all and the charter is signed 'off-stage': the crucial act itself is elided in the stanza break between the imperative and graphologically emphatic Barons' '"Sign!"' and the penultimate verse in which 'Your rights were [already] won'. To textualise the alternative, seeing John directly sign, would have been to assign him too much positive agency, and so he and his actions are negated away.

Lacuna effects are most obvious in stylistic negations like these, but it is also worth considering the conceptual mechanics of lacunae in different stylistic forms. Text world theory (Werth, 1999; Gavins, 2007) neatly explains the lacunal sense of felt absence by modelling inferencing (like Iser, 2000) as world-building: a reader fills out a rich conceptual text world using both the features in the text and their own experience. Any ontological shift away from the ongoing text world during a reading serves to create a 'world-switch' that creates another, similarly rich world which is subordinate or alternate to the main initial text world. Negation is one trigger for such a world-switch. In the poem's 'never break', for example, the text world of the reeds at Runnymede contains a brief switch to a sub-world in which the reeds break, and this world is related to the main text world negationally. The model thus neatly captures the feeling of brokenness while at the same time maintaining the denotational meaning of the text. Similarly, the first direct speech contains a negation 'You mustn't sell . . .', which again evokes a brief world-switch to a scenario in which rights and liberties *have been* sold, only for the main text world to reassert itself: the lurking dangerous alternative is thus evoked at the same time. In this particular example, the switch is even more complex as the negation is embedded in a deontic modal 'must', and modality is another world-switch trigger (Gavins, 2005). Here the switch is an ethical one – the compared difference between what is and what must be – but again the world-switched conceptual mechanics explains how both the prohibition and the danger can be held in mind in parallel.

Later in the poem, after the signing of the charter, the negations and modals combine again with a different effect. In 'No freeman shall be fined or bound', the effect of using the negative particle rather than a grammatical negation (as in 'Freemen shall not be fined or bound') is to create a primary negation followed by a secondary modal. This conceptual world-sequencing then allows the grammar after the initial 'No' to be oriented positively, authoritatively and with prophetic certainty. The sense is more emphatic and (ironically, given the charter) less legalistic in tone at the point at which it follows the highly emphatic 'Your rights were won at Runnymede!' However, because the charter *is* a legal document, the syntax that follows switches back again into the switched world (cued by 'Except') where fining, binding and dispossession happen 'by lawful judgment'. The stanza maintains the balance between the stirring main event (at text world level) and its calm legal statement (at sub-world level). The stanza ends by making the negation secondary again ('Forget not', rather than 'Do not forget'), in order to confer attentional prominence on the sub-world possible danger of forgetting: this neatly sets up the sense of risk and required vigilance that is the theme of the final stanza.

Other stylistic triggers of world-switches include metaphor and direct speech, both of which also feature significantly and inter-relatedly in the poem. The most obvious metaphors are the personifications of the reeds and the river Thames, enacted very prominently in the opening stanza: 'What say the reeds . . . the lissom reeds . . . they keep the sleepy Thames awake with tales'. In cognitive grammatical terms (Langacker, 1990, 2008), personifications constitute a highly prominent shift of role from abstractions and inanimates to focused figures; a figure makes a strong 'attractor' of attention if it also moves, speaks and has agency, as here do the reeds (Stockwell, 2009a: 25). The Thames is also personified, but at this stage the activity and empathetic role is not as strong as the reeds: the Thames is in patient rather than agent position, rheme rather than topic, and is sleepy rather than active and restless. (The Thames in the final stanza is different in these respects.)

Connected to the personification of the reeds is their direct speech, which fills the central part of the poem from the middle of the second stanza to the end of the penultimate verse. The attentional distraction effected by the personification metaphor and the direct speech switch serve to occlude the speaking poetic voice who initially asks, 'What say the reeds . . .' The various world-switches I have outlined so far all combine to deflect the reader's attention into the poem, drawing the viewpoint in to the historical and romanticised event in the distant past, and making it immediate, personal and direct (at the discourse-world level which is occupied by the reader and the notion of the author, Kipling). The generalised 'You' of the second stanza becomes a personal address ('Your rights') as you are drawn into the scene.

There is a balanced symmetry created as a result, where Kipling's authorial voice in the same modern discourse world as the reader begins in written, documentary form (in the sub-title, in the association of reeds with writing implements, and perhaps even in the homophonic potential of 'reeds'/ reads), and then moves into self-conscious poetic forms ('What say . . .', 'lissom', 'with tales', 'Oh, hear' – this last possibly recalling for readers of a literary bent the same repeated phrase from Shelley's revolutionary 'Ode to the West Wind'). The reeds then speak as the people of England, in a relatively conversational syntax filled with exclamatory markers and demotic phrasing ('saw 'em roused', and 'settled John'). The speaking reeds shift the focus from writing to speech, until the very centre of the poem and a further piece of embedded direct speech which in fact denotes writing – '"Sign!"' – a word that itself draws attention to its own symbolic value. The penultimate stanza maintains the speaking reeds' voice, ending with a prefiguring echo of writing ('signed at Runnymede'). The final stanza returns in balance to the top world level of the Kipling voice, again with a highly self-conscious poetic register: the same heavy alliteration as in the first stanza, the same highly-wrought metonymy 'when Mob or Monarch lays too rude a hand'. This time, though, the focus is on the spoken voice rather than on writing ('whisper', 'shudder', 'warning'), and it is the Thames who is now woken, vigilant and knowing, fully empathetic with 'the moods of kings,/And crowds and priests and suchlike things'. This combination enlarges the Thames and makes him all-encompassing, makes him heavy, noisy, personified, speaking, active and definite – all these aligned as highly distracting attentional features.

The final stanza also creates an onward rolling trajectory at the end of the poem, and it is perhaps this dynamism that makes the final two stanzas the most quotable in popular usage. Previously, the reeds have bent but not broken, have spoken but mainly observed, and the Barons only 'lumbered' slowly. Runnymede itself is static, serving as the initial background and final grounding of each sentence and stanza and following the fixing preposition 'at'. The final stanza begins 'still', but then there is activity and motion: hands being laid, waking, playing, shuddering, and a moving preposition 'Across the reeds'; and then the moving river Thames rolling both literally and figuratively in his voice and figuratively too through history, ending with two different prepositions 'down from' Runnymede to indicate the trajectory even more prominently. There are some balancing parallels between first and final stanzas, then, but these also create a background that points up the crucial differences that the signing of the charter in between has created. Similarly at a phonological level, the heavy alliteration that is a feature of both first and final stanza is matched by a shift from the predominantly riverly and reedy liquids, sibilants and voiceless sounds of the opening to a greater preponderance of voiced portentous sounds in the final lines.

The final type of world-switch in evidence in the poem is the temporal shift into the past with the reeds' direct speech; the poem ends with a jump in the final stanza to the continuing present ('still when Mob or Monarch lays'), followed by a hint of vigilance, unspoken threat and unspecified power in the future ('deep and dreadful as he brings their warning'). The intensity of this final admonition takes much of its force from the resonance of all the lacunal moments throughout the poem, all of the world-switches that have evoked the modalised and negated glimpses of evil practice, undesirable situations and worse possibilities. They are hinted at but unsaid in the final stanza, the trajectory of the syntax, time, the river, the attentional shifts and the implied exit motion between 'at' and 'down from', all pointing to an unexpressed but powerful ongoing defence of liberty outwards into the readerly discourse world after the poem has finished.

The foregoing has been mainly concerned with the internal, text-driven lacunae that readers need to process creatively. Of course, there are in addition a great deal of framing silences that can also be regarded as lacunal absences by a particularly disposed reader. The poem's version of Magna Carta is one of universal liberty, but this leaves unarticulated the fact that only landowners, and landowning men, and white English landowning men were to possess these rights until the early twentieth century. The poem presents an eternal vision of Englishness, though in 1215 both the barons and the king were essentially still Norman French, and the English as a people were less than 400 years old. The Kipling voice of late nineteenth century British imperialism avoids the trampling of the rights of black Africans, trade unionists, chartists and working people whose liberties were indeed sold, delayed and denied, and it restricts the values of the British empire and people implicitly to the English. The conflicting interests of monarchy, state, people and church are idealised as a balance that can be maintained by law rather than reconfigured more fairly by struggles either violent or political. These are all also matters of the creative imagining of the poem, rather than irrelevant matters to be boxed away as 'context', for those readers appropriately disposed to enact such roles in relation to the poem. Crucially, although earlier I distinguished internal indeterminacies from external silences, in fact there is no readerly difference between the two: both are interanimating facets of a reader's creative fulfilling of the text, and both aspects of reading can be understood in the same consistent cognitive poetic terms.

Even for a reader antipathetic to a right-wing, imperialist or extreme libertarian position, and though the attraction of the text to conservative US Republicans or Thatcherite English jingoists might leave a bad taste, there is no question that many readers even of a more liberal or socialist inclination still feel a frisson of resonant power in the Kipling poem. This is partly because it is almost impossible to read it in an engaged way while simultaneously being resistant to it: it assembles a certain sort of reader very strongly, largely through the mechanics of its world-switch construction (most resistance comes after the reading, not during it). The alignment of negations and other world-switches with the alliterative and other phonological patterns prompts the reader to create resonant lacunae that produce a rich, multi-layered world, far fuller and more tangible than the apparent closedness and relative determinacy of the text might suggest. The poem's apparent balancing of elements as set out above serves to distract a reader from all that is elided or unsaid. There is a seduction in the aesthetic stylistic arrangement of the poem that extends to the relationships between Kipling, reader, reeds, river and people in the world, and these aesthetic effects have a bearing on the ethical position that a reader is encouraged to take inside the poem. This temporary, soft-assembled and internal readerly entity might be set aside as the person sets aside the poem after reading it, but I suggest even so that there is a persistent, resonant trace left in that person's life. Crucially, here, all these matters can be theorised and explored in ways that are psychologically plausible and consistent with the reader's engagement with the stylistics of the text.

As cognitive poetics develops as a field, we are finding that poetics, aesthetics and ethics are intimately bound together, and the distinctions between emotional and rational responses to creative verbal art are increasingly difficult to sustain coherently (see work by Phelan, 2005, 2007). The literary reading of information, feeling and value are interanimating, and this requires a theoretical and practical account of imaginative creativity that is offered by the continuities of cognitive poetics. We are coming to a position where we can better understand how we feel about knowledge and beliefs, and how creative the reading process really is.

References

Attridge, D. (2004) *The singularity of literature*. London: Routledge.

Balibar, E. and Macherey, P. (1978) 'On literature as an ideological form' (trans. McLeod, I.). *Oxford Literary Review* **3** 4–12.

Bloom, H. (ed.) (2004) *Rudyard Kipling*. Philadelphia: Chelsea House.

Buchanan, R. (1899) 'The voice of "The Hooligan"'. *Contemporary Review* **76** (Dec.) 774–89.

Bucholtz, M. and Hall, K. (2005) 'Identity and interaction: A sociocultural Linguistic Approach'. *Discourse Studies* **7** (4–5) 585–614.

Carstensen, K. U. (2007) 'Spatio-temporal ontologies and attention'. *Spatial Cognition and Computation* **7** (1) 13–32.

Eco, U. (1972) *The open work* (trans. Cancogni, A. from *Opera Aperta*, 1962). Cambridge: Harvard University Press.

Eco, U. (1979) *The role of the reader: Explorations in the semiotics of texts*. Bloomington: Indiana University Press.

Fletcher, C. R. L. and Kipling, R. (1911) *A school history of England*. Garden City, New York: Doubleday, Page & Co.

Gadamer, H. G. (1989) *Truth and method* (trans. Weinsheimer, J. and Marshall, D. G. from *Wahrheit und Methode*, 1960) (2nd edn). New York: Crossroad Press.

Gavins, J. (2005) '(Re)thinking modality: A text-world perspective'. *Journal of Literary Semantics* **34** (2) 79–93.

Gavins, J. (2007) *Text world theory*. Edinburgh: Edinburgh University Press.

Gibbs, R. W. (2003) 'Prototypes in dynamic meaning construal' in *Cognitive poetics in practice*, Gavins, J. and Steen, G. (eds). London: Routledge, 27–40.

Gilmour, D. (2003) *The long recessional: The imperial life of Rudyard Kipling*. New York: Farrar, Straus and Giroux.

Ingarden, R. (1973a) *The literary work of art: An investigation on the borderlines of ontology, logic, and theory of literature* (trans. Grabowics, G. from *Das literarische Kunstwerk*, 1931). Evanston: Northwestern University Press.

Ingarden, R. (1973b) *The cognition of the literary work of art* (trans. Crowley, A. R. and Olson, K., from the German *Vom Erkennen des literarischen Kunstwerks*, 1968; original Polish *O poznawaniu dziela literackiego*, 1937). Evanston: Northwestern University Press.

Iser, W. (1974) *The implied reader: Patterns of communication in prose fiction from Bunyan to Beckett*. Baltimore: Johns Hopkins University Press.

Iser, W. (1978) *The act of reading: A theory of aesthetic response*. Baltimore: Johns Hopkins University Press.

Iser, W. (2000) *The range of interpretation.* New York: Columbia University Press.

Lakoff, G. (1987) *Women, fire and dangerous things: What categories reveal about the mind.* Chicago: University of Chicago Press.

Lakoff, G. (2002) *Moral politics* (2nd edn). Chicago: Chicago University Press.

Langacker, R. (1990) *Concept, image and symbol: The cognitive basis of grammar.* Berlin: Mouton de Gruyter.

Langacker, R. (2008) *Cognitive grammar: A basic introduction.* Oxford: Oxford University Press.

Macherey, P. (1978) *A theory of literary production* (trans. Wall, G. from *Pour une théorie de la production littéraire,* 1966). London: Routledge and Kegan Paul.

Macherey, P. (1985) *The object of literature* (trans. Macey, D. from *A quoi pense la littérature?* 1990). Cambridge: Cambridge University Press.

Mallett, P. (2003) *Rudyard Kipling: A literary life.* Houndmills: Palgrave Macmillan.

Peterson, M. A. and Enns, J. T. (2005) 'The edge complex: Implicit memory for figure assignment in shape perception'. *Perception and Psychophysics* **67** (4) 727–40.

Phelan, J. (2005) *Living to tell about it: A rhetoric and ethics of character narration.* Ithaca: Cornell University Press.

Phelan, J. (2007) *Experiencing fiction: Judgments, progressions, and the rhetorical theory of narrative.* Columbus: Ohio State University Press.

Smith, C. W., Johnson, S. P. and Spelke, E. S. (2003) 'Motion and edge sensitivity in perception of object unity'. *Cognitive Psychology* **46** 31–64.

Stockwell, P. (2002) *Cognitive poetics: An introduction.* London: Routledge.

Stockwell, P. (2003) 'Surreal figures' in *Cognitive poetics in practice,* Gavins, J. and Steen, G. (eds). London: Routledge, 13–25.

Stockwell, P. (2005) 'Texture and identification'. *European Journal of English Studies* **9** (2) 143–53.

Stockwell, P. (2009a) *Texture: A cognitive aesthetics of reading.* Edinburgh: Edinburgh University Press.

Stockwell, P. (2009b) 'The cognitive poetics of literary resonance'. *Language and Cognition* **1** (1) 25–44.

Stockwell, P. (2010) 'Accommodating identity in cognitive sociolinguistics'. *Cahiers du CEL* 14. (Lyon: Centre d'Etudes Linguistiques, Université Jean Moulin.

Sullivan, Z. T. (1993) *Narratives of empire: The fictions of Rudyard Kipling*. Cambridge: Cambridge University Press.

van Dijk, T. (1980) *Macrostructures: An interdisciplinary study of global structures in discourse, interaction, and cognition*. Hillsdale: Erlbaum.

van Dijk, T. (2008) *Discourse and context: A sociocognitive approach*. Cambridge: Cambridge University Press.

Werth, P. (1999) *Text worlds: Representing conceptual space in discourse*. Harlow: Longman.

CHAPTER 4

Creative interpretations: Discourse analysis and literary reading

Joan Swann

Introduction: a reader-oriented approach to creativity

> A clear requirement now is to embrace not simply the producer but the *receiver* of creative processes . . .
>
> (Carter, 2007: 600)

Ronald Carter's comment above is a response to a special issue of the journal *Applied Linguistics* on 'Language creativity in everyday contexts' (Swann and Maybin, 2007). The special issue took as its starting point a significant strand of research, including Carter's own work, that has focused on the pervasiveness of everyday linguistic creativity. 'Creativity' is associated here with literary, or literary-like, language, and an argument emerging from several studies is that features associated with literary texts (poetic forms such as word play, figurative language, repetition or parallelism; story and the creation of imaginative worlds) are routinely found in more everyday contexts. Researchers have explored the range of forms such literary-like behaviour may take; its interactional and interpersonal functions; and potential cognitive, and even evolutionary benefits (Carter, 1999, 2004; Cook, 2000; Crystal, 1998; Goddard, 2003; Maybin and Swann, 2006; Norrick, 2000/2001; Tannen, 1989/2007).

The special issue sought to extend this earlier research in developing a more socially oriented, contextualised and critical understanding of everyday linguistic creativity. Carter's argument was, however, that papers still focused predominantly on production, and that greater account needed to be taken of reception, 'with the aim of gaining enriched understanding of processes of reception on the part of different socially positioned readers or viewers of or participants in creative performances'. The present chapter represents a further turn in the dialogue, and begins to address the 'reception' gap identified by Carter.

A great deal of existing research in language creativity – including Carter's own corpus-based study – has focused on various forms of interaction, spoken or online. In this case, creative episodes such as poetic language, conversational narrative or other forms of interactional performance are co-constructed between participants, so that it is not always possible to sustain straightforward distinctions between 'production' and 'reception'. Reception is however somewhat easier to address as a distinctive process when it is separated from earlier processes of production, as in the case of reading literary or other texts. Carter himself notes that the research he envisages would parallel developments in literary criticism that have led to calls for more studies of reading and reader response.

Below I illustrate responses to literature evident in the discourse of contemporary reading groups. Reading groups are of interest as a widespread literary practice. Since they come together explicitly to talk about books, they provide a good opportunity for the discourse analyst to study literary reception, or one type of literary reception, on the hoof in particular reading contexts. While literary texts are (archetypally) creative, and so their reception is relevant to the study of creativity, I shall suggest also that literary discussion is discursively creative in that reading groups collaboratively construct textual meanings or interpretations. As a background to this discourse analytic approach, I first consider some contrasting approaches to the study of literary reading and readers.

Studying literary reading

On the face of it, it may seem odd to call for more studies of reading/reception. Ideas of 'the reader' are prevalent in literary theory, and the reader is also routinely appealed to in scholarly interpretations of particular texts. What is at issue here, however, tends to be an abstract and idealised[1] reader rather than the socially positioned reader envisaged by Carter. I shall provide a brief illustration from literary stylistics, a tradition of literary criticism that draws on linguistic and/or discourse analysis to explore the meaning of literary texts.

In his well-known introduction to literary stylistics, Mick Short argues that 'stylistic analysis, which attempts to relate linguistic description to interpretation, is part of the essential core of good criticism' (1996: 5). A benefit of close textual analysis is that it can help to resolve any differences in potential meanings ascribed to a text. At interest, however, is not simply what a text means, but also how it comes to mean what it does: '. . . we are profoundly interested in the *rules and procedures* which we, as readers, intuitively know and apply in order to understand what we read.' (1996: 6; italics as in original). As an illustration of how texts come to mean, Short gives the example of linguistic deviation (where an utterance doesn't conform to expected textual patterns): this is associated with foregrounding, a cognitive effect whereby a textual element (e.g. part of a poem) becomes noticeable or 'perceptually prominent' (p.13).

Both what a text means, and inferential procedures adopted by readers, are derived here from textual analysis (rather than, say, from consultation with actual readers).[2] Short concedes there may be differences between readers but focuses on an assumed significant amount of agreement on textual meaning. Context is invoked ('to understand [a] sentence completely we also have to see it in its context,' p. 8) but fairly narrowly interpreted. In the case of a line of dramatic dialogue, for instance, events in the play, along with our 'general world knowledge', will lead to a sentence being interpreted in one way and not another.[3] In this volume, Michael Toolan's contrast between poems by Seamus Heaney and William McGonagall represents a similar stylistic tradition, in which readers and reading positions are explicitly invoked on the basis of textual analysis.

Recent years have seen a 'cognitive turn' (Stockwell, 2002) in literary stylistics, and potentially in literary study more generally, with the advent of cognitive poetics. Cognitive poetics is a more explicitly reader-oriented approach to the study of literature, drawing on specific cognitive theories to model interpretive processes engaged in by readers. Peter Stockwell's introduction to this relatively new field argues that it offers 'a unified explanation of both individual interpretations as well as interpretations that are shared by a group, community or culture', and that, in its focus on reading, 'it represents nothing less than a democratisation of literary study' (2002: 5, 11). Despite this conceptual refocusing, many studies restrict themselves to textual analysis, with an appeal to an abstract and idealised reader that looks remarkably similar to traditional stylistic analysis. Stockwell's analysis of Kipling's 'The Reeds of Runnymede' in this volume represents an illustration. Some cognitively oriented studies do go further, however, drawing on experimental evidence from the interpretations of actual rather than idealised readers.

If cognitive poetics represents a cognitive turn in literary research, then the study of actual readers' engagement with texts may be seen as a further empirical turn. In a review of what they term 'the empirical study of literature', David Miall and Don Kuiken (1998) and Miall (2006) raise great hopes for this approach. Whereas literature scholars, they argue, 'produce readings of texts and elaborations of literary theory in an institutional culture that is inhabited almost exclusively by fellow scholars and senior students' (1998: 328), an empirical approach concerns itself with the ordinary reader, outside this institutional culture. Miall and Kuiken are interested, then, in everyday interpretations of literary texts, arguing that this will produce 'a more ecologically valid approach to understanding the role and functions of literature in general' (1998: 328).

Miall and Kuiken's work remains textual, in its focus on the formal properties of literary texts (in this sense it is similar to much of stylistics), but the intention is to examine these 'in cooperation with actual readers'. As an example they discuss a study of foregrounding in which readers (students with different levels of experience of literature) were asked to read short stories by Woolf and Mansfield. Segments of the stories were coded for certain foregrounded features, and readers' responses to these were measured – e.g. they rated segments in terms of characteristics such as 'affect' and 'strikingness'. Miall and Kuiken found that both experienced and inexperienced readers responded to foregrounding, but that experienced readers gave higher affect and strikingness ratings than those with less literary experience.

In its focus on real readers, research in the 'empirical' tradition seeks to add a new dimension to the study of literature. An important limitation however is that, as in the example above, studies have restricted themselves by and large to an experimental paradigm, often using students as subjects. While this allows the exploration of readers' responses to specified textual features, the conditions under which reading occurs are necessarily artificial and the reading behaviour itself is atypical. This throws into question Miall and Kuiken's claim to ecological validity, and the potential of their work to provide general understandings of the role and functions of literature (cf. Allington and Swann, 2009; Hall, 2008).

More contextualised accounts of literary reading are found in two rather different established bodies of work: the historical study of reading experiences based on accounts in diaries, marginalia, letters, autobiographies etc. (e.g. Rose, 2002; see also discussion in Halsey, 2009); and contemporary ethnographic studies of popular and canonical fictional reading (Cherland, 1994; Radway, 1984/1991; Zubair, 2003), including Elizabeth Long's study (2003) of women's reading groups. Geoff Hall characterises such studies under the broad heading of literature as social practice, but argues that research is still limited and that further work is needed in this area:

> . . . we need to study people's reading and writing activities in the broader contexts of their lives and social interactions. The empirical investigation of what people do with texts and how they find value in literature within the constraints of their own social contexts has been neglected.
>
> (Hall, 2006: 451)

Historical and ethnographic studies provide a focus on contexts of reception and see readers themselves as historically and socially positioned, but they do not attend, at any level of detail, to specific interpretive processes. This perspective may, however, be provided by a discourse analytic approach, seen in a relatively small number of recent studies that have taken literary discussion as their primary object of study (e.g. Allington (2007) on slash fiction; Benwell (2009) on the reception of diasporic fiction; Eriksson and Aronsson (2004, 2009) on school 'booktalk'; Swann and Allington (2009) on the interpretive activity of reading groups). A discourse analytic approach would share with historical and ethnographic research an interest in contextualised reading practices. In its focus on the interpretive processes engaged in by readers it also has something in common with the rather different intellectual traditions of cognitive poetics and the empirical study of literature, although its concern with the local discursive construction of readings differentiates it methodologically from these traditions.

Reading is often viewed as a silent and solitary activity, in which interpretation occurs in the interaction between an individual reader and a text. However, people also talk about their reading in a wide range of contexts – bedtime stories and other joint reading activities, informal conversation between friends, media discussion, academic seminars, literature classes, public meetings between readers and authors, and reading groups. Such talk may be seen as an interpretive activity, or a readerly act. In reading group talk, for instance, readers do not simply report on their earlier private reading. Reading group members comment that, through discussion, their interpretations may be changed, or perhaps extended (Allington and Swann, in press, 2011). Discourse analysis of discussions reveals a more complex process in which interpretations are collaboratively and creatively co-constructed.

This kind of literary talk, therefore, provides an important source of evidence on how readers interpret books and, for the discourse analyst, an opportunity to study this as a discursive phenomenon. Below, I provide an example of this process at work in contemporary British reading groups.

An analysis of literary discourse

The data I discuss below come from a project entitled *The discourse of reading groups*,[4] funded by the British Arts and Humanities Research Council. The project is designed to investigate the particular kinds of interpretations and evaluations reading group members provide of literary texts and how these interpretations/evaluations are discursively constructed in reading group interaction. Our sample of 16 face-to-face groups was designed to reflect some of the diversity of contemporary reading groups: groups come from different parts of Britain; they include a mix of age ranges, and both single-sex and mixed-sex groups; they meet in different locations (in members' houses, in a pub and a gay bar, in bookshops, a library, a community centre, a school, a higher education department and a prison); there are different types of relationship between members (e.g. friendship, various kinds of institutional relationship).

The project data consist of audio-recordings and field notes from group discussions; audio-recorded informal interviews with group members, and sometimes follow-up email discussion; and documentary evidence such as press reviews and published criticism of books read by groups, newsletters, websites and book reviews written by members. We have adopted various analytical tools in our data analysis, including qualitative analytical software (Atlas-ti) and corpus linguistics (WMatrix). In this chapter, however, I draw on a qualitative form of discourse analysis derived from interactional sociolinguistics (Gumperz, 1982) and incorporating insights from other sociolinguistic studies of talk in both friendship contexts (e.g. Tannen, 1984/2005; Coates, 1996) and institutional contexts (e.g. Holmes and Stubbe, 2003; Swann, 2007). Such analysis is sensitive to the social and interactional context in which talk is produced, although 'context' is seen not simply as a phenomenon that stands outside the interaction but also as something that is dialogically constructed between participants. The implications of this relatively contextualised approach for the understanding of literary reading are addressed further below. Within this broad approach, analysis focuses both on the identification of emergent themes and the discourse strategies adopted by participants in constructing interpretations of literary texts.

To illustrate the potential of this form of analysis I shall examine data from two contrasting reading groups. Both groups are responding to the same book: John Steinbeck's *The Pearl*, a novella concerned with a poor family of pearl divers who find a valuable pearl that seems, at first, to offer the hope of great wealth. One group, which I refer to as the medics group, is an interprofessional group of staff and students in a university medical faculty. The reading group provides an opportunity for colleagues from different areas to meet and discuss, and is seen by one of its conveners as contributing to staff development. Meetings are held at lunchtime; some members bring their lunch, and hot drinks and water are available in the meeting room. There are two conveners, both members of staff, who also act as facilitators. While some members attend regularly, membership is not fixed. Six members attended the meeting I discuss below. The second group has in common with the first the fact that its members also work in a university. However, this is a group of 11 women friends who don't have a professional connection. One member took the initiative in starting the group, but they do not have a facilitator or any similar formal role. The group meets in the evening in one another's houses. The host serves drinks, including wine, and snacks, and other members also bring food to share. Food is sometimes themed to fit the book. Members noted they had difficulty in this case with *The Pearl*, but some brought tortillas or nachos.

In general terms, the groups responded rather differently to the novella. The medics were mainly positive, whereas the friendship group disliked the book, though one member commented favourably on the writing style. Some themes recurred in each discussion – e.g. characterisation and the relationship between characters; the moral or message in the story; writing style including the use of imagery. Group members took different positions on these, however: for instance, the friendship group spent more time discussing characters and commented more negatively on characterisation, particularly on the portrayal of the main character's wife. Below I look more closely at how these discussions work, focusing in turn on the relationship between literary reading and readers' personal experiences and the collaborative co-construction of readings.

Literary reading and personal experience

In the medics' group, a connection is often drawn between literature and medicine. Two students, Jon and Rachel, refer to a Special Study Module they are taking as part of their course (with the reading group facilitator, Meg) on the use of literature in medicine. Group members are invited to bring poems to the meeting, and Jon brings a humorous poem by Adam Kay: 'A letter to the patient's GP please Angela'. In discussing whether to read something by Ian McEwan in the future, Jon comments that *Saturday* 'really solidified my desire to become . . . a medical professional', and Rachel that: 'I've been told by quite a few medics that *Saturday*'s the one to read'.

Such experiences are also drawn on in discussion of *The Pearl*. For instance, the facilitator, Meg, refers to plans made by the main character, Kino, after he finds the pearl, reading out a sentence from the text: 'A plan is a real thing, and things projected are experienced.' Meg relates this idea to becoming a doctor:

> Seeing yourself as a doctor you know the more you can picture yourself (.) in that role the more the more real it becomes (.) if you if you (go around xxx) **((mock groaning noise))** you know **((laugh))** it's never gonna happen **((laugh))**

In this case the connection between literary text and readers is drawn explicitly: a generalised, or potentially generalisable comment derived from events in the text becomes recontextualised as a moral or lesson for medical students.

One of the characters in the novella is a doctor, and this provides a further stimulus for discussion of real-world experiences.

Extract 4.1 Doctors, knowledge and power

The group has been discussing the oppression faced by Kino's family and other poor villagers, and as an example Meg refers to the corrupt doctor in the village, who refuses to treat Kino's baby for a scorpion sting until he learns about the valuable pearl. At this point he visits the family and administers a powder that makes the baby appear ill so that he can claim a fee for treating him.

1	Meg:	But that's wh- that's what happened when he went to the doctor wasn't it (it) was the doctor had this (.) power and-
2	Jon:	I didn't like how they port[rayed doctors
3	Meg:	[and it says page
4	Rachel:	yeah ((slight laughter))
5	Meg:	page 33 you know when the when the doctor is saying oh no you know the doctor has come back and said [oh I
6	Rachel:	[Yeah when

7	Meg:	[need t-
8	Rachel:	[he comes back
9	Meg:	oh yes so I can he- you know I can give your <mmh> child this medicine and I can you know <mmh> em ((reads from book)) 'and Kino felt the rage and hatred melting towards fear he did not know and perhaps this doctor did <mmh> and he could not take the chance of pitting his certain ignorance against this man's possible knowledge <mmh> he was trapped as his people were always trapped and would be until as he had said they could be sure that the things in the books were really in the books
10	Rachel:	yeah (.) knowledge and power
11	Jon	something that doctors have over patients isn't it the the the power of knowledge about a problem of the patient's
12	Meg:	yeah and that kind of
13	Rachel:	((quiet)) helplessness
14	Meg:	special knowledge <mmh> you know but
15	Jon	that Google is er- eroding away ((laughing voice))
16		((laughter))
17	Meg:	yeah the relationship is changing yes <mmh> but that that to me that was what it was about was that that would give his son a chance to ((knocks on table))
18		<Yeah mmh>
19	Beth	but I think it is quite difficult cos you can't I found it really difficult to try and work out a time and a place . . .

For the transcription conventions used in this and other extracts please see the list at the end of the chapter.

In the first turn (1), Meg refers to the doctor's power. She develops this point, citing (9) a passage from the novella that refers to Kino's inability to challenge the doctor because of his likely knowledge and Kino's ignorance. Jon's comment (2) that he doesn't like the portrayal of doctors receives interactional support from Rachel (4) but when Meg continues speaking it is not developed further here.

Meg's point is developed with interactional support from Rachel, in her partial repetition of Meg's utterance ('yeah when he comes back' (6 and 8) and her later recycling of the two terms 'knowledge' and 'power' (10). This is elaborated on by Jon and recontextualised as a general statement about doctors' power over patients (11), in turn elaborated on by Meg (12 and 14). Rachel's potential completion of Meg's turn ('and that kind of/helplessness') is uttered quietly and not acknowledged. Jon's reference to Google eroding doctors' power (16) is uttered in laughing voice and responded to with laughter from others. It is syntactically ambiguous, potentially continuing Meg's turn (14), his own prior turn (11), or both. Meg offers further agreement and support ('yeah the relationship is changing yes'). Meg then returns to an earlier point in the discussion (Kino's desire for an education for his son) and Beth introduces a new topic – the difficulty of identifying time and place in the novel (19).

Here, then, the doctor's power in the novella is related to contemporary events within readers' experience. The development of this interpretation is also associated with some interpersonal activity. For instance, Jon seeks to distance himself from the portrayal of the doctor. His later acknowledgement of doctors' knowledge and power is made in the third person – again perhaps a distancing device. The laughing voice that accompanies his reference to Google's erosion of medical power turns this into a slightly humorous statement – maybe a wry reflection on something that will affect him as a practitioner. At any rate the humour works, responded to by laughter from others. Humour and laughter throughout the discussion contribute towards the maintenance of friendly relations in a group that generally plays down potential hierarchies between participants.

The friendship group also relates their discussion of the book to personal and 'real world' experiences. For instance, this is a women's friendship group and their identity as women is sometimes made salient. Readers often comment on the condition of women characters, and when asked about this later they say this is an aspect of 'who we are' and 'a factor of the interests of some of the group'. One member comments in a follow-up email:

> We don't all share the same perspectives but we do comment a lot on things like whether the author creates good women characters, whether they represent experiences we understand or empathise with. We tend to have strong feelings about this . . .

An example from discussion of *The Pearl* is a criticism of the portrayal of Juana, Kino's wife.

Extract 4.2 Reading as a woman

The group is talking about their dislike of the book. Another reader has commented that it is too black and white, and Clare agrees, giving examples. She is also critical of the author's depiction of characters, then turns to her own reading position as a woman:

1 Clare: Needless to say it's difficult to be a woman and read this I think <mmh>
I mean <mmh> the portrayal of the wife
[was kind of

2 Helen: [yes it's so (xxx) ((laughing voice))

3 Clare: close to nature and [(.) mhh in some way <yes> (.)

4 Helen: [yes ((laughter))

5 Clare: intrinsically <mmh> (.) good <yes> and the l- the sort of
(light house) the sort of register of pure truth against which the man
will you know deviate but he will come back and she will still be there
however badly he treats her as his <yeah> wife <yes> all the rest of it
I mean you know I can do without that <mmh> thank you very much
and the thought of teenagers actually being fed <yes> that kind of
<yes> view of life <mmh> is is is is ki- would be <yeah> worrying

Clare here comments on the difficulty of reading the novel as a woman, and relates this particularly to the portrayal of Juana as close to nature, intrinsically good, representing 'pure truth' etc. There has been some earlier discussion of the book as a text that is 'put up as a moral tale for teenagers' and seen as 'kind of a classic for children' and Clare returns to this point here: the thought of teenagers being fed this view of life is worrying. Clare receives some interactional support from Helen and there is evidence of assent from others in a relatively high level of minimal responses. Helen's laughter may again serve as a wry reflection on the stereotypical portrayal of Juana.

The collaborative development of literary readings

The extracts above show that, in these contexts, literary reading is a discursive accomplishment in which readers collaboratively construct interpretations. This is more developed in the longer sequence in Extract 4.1 – for instance in participants supporting, building on and elaborating one another's contributions. Extract 4.3 below, from the friendship group, shows how, in the sequential development of readings, readers may collaboratively arrive at a more complex interpretation of an aspect of the narrative.

Extract 4.3 No turning back

Becky, who did not like the book, is referring to one or two places where she feels, more positively, that the author is telling her something she hadn't thought of. She refers first to a comment in the book that humans are never satisfied with what they have, and that though this is disparaged it is actually one of the greatest talents of the species. Below she turns to a second point:

	Speaker	Utterance	Comment
1	Becky:	. . . and then page fifty-five was the only other thing (.) em when he when he went away and lost his old <mmh> life I thought that was very well <yes> expressed never in his life he had been so far from home he was terrified with all these things that lay over the water over the mountains and he lost his old world and he must clamber on to a new one (.) and there is no turning back that feeling that there's no turning ba- once you have knowledge you can't undo knowledge and you can't undo [experience	slow, measured; close paraphrase from book, some phrases verbatim
2	Jen:	[(and yet) he did	
3		(2.0)	
4	?	He [turned back	
5	Becky:	[And yet he did	Question intonation
6	Jen:	He turned back he went back	
7	Becky:	He did turn back	Slight emphasis; acknowledgement
8	Jen:	So you know those kinds of statements	
9	Becky:	[yes	
10	Jen:	[were grander than the story 'cos you know the [first one about the great thing it is	
11	Becky:	[yeah	
12	Jen:	is that the species is not is not satisfied <mmh> with what it's got it wants more and that's its great thing and yet the whole story seems to be saying you are much better off being satisfied with what <yes> you've got not trying to improve <mmh> yourself cause otherwise your baby gets its head shot off (.) <yeah> so you know ((slight laughter)) the same with it <mmh> there is no turning back and what's the end of the story he's back [in the village	some unclear overlapping speech at this point
13	Becky:	[he's back	
14	Helen:	He turned back but he could never go back to his simple life because of what had happened to him so tha- in that sense	
15	Becky:	in that sense [mmh	
16	Jen:	[he went back but he was a different person yeah <mmh yeah>	

Becky refers to an episode in the narrative where Kino has rejected an unreasonable offer for his pearl from the pearl buyers who visit the village. He must now leave the village and try to sell the pearl in the capital. The episode marks the loss of Kino's old life along with his fear of the new. Becky comments on the more general lesson that you cannot undo knowledge and experience – there is no turning back from this. Jen offers a contradictory view – 'and yet he did' (2), referring to Kino's eventual return to the village. After a two-second pause (noticeably long in this interaction) another reader elaborates Jen's utterance: 'He turned back'. Becky's repetition (5) is a question to Jen, who further elaborates her earlier comment (6). Becky acknowledges this and later agrees with Jen's further points that certain general statements are 'grander' than the events depicted in the novella. However, Helen then presents an alternative view, or at least a qualification – while Kino returned to the village he could not go back to his simple life. This is accepted by Becky in her repetition of 'in that sense' (15), and by Jen who also extends the point – 'he [Kino] was a different person' (16).

The extract illustrates the sequential development of a brief literary argument, as an original interpretation of the novella is challenged, an interim consensus reached between some readers, which is in turn qualified leading to a new consensus. While specific argumentation patterns differ, throughout our sample people routinely build on one another's ideas to reach new understandings of their reading. In this case Helen bridges Becky's and Jen's earlier points – Kino turned back in one sense but not in another – to construct a more sophisticated interpretation of an aspect of the narrative.

A noticeable interpersonal feature in this episode is the lack of mitigation in managing disagreement. This is rare in our data – more usually challenges are hedged as participants attend to one another's face. Lower levels of mitigation seem to be associated with one or two close friendship groups, where members are able to absorb bald challenges while sustaining a high level of collaboration. This is remarked on later by some members of the friendship group above in their responses to questions about the nature of their discussion – e.g.:

> after almost five years we know each other pretty well, so any inhibitions have disappeared!
>
> (email correspondence)

Discussion

In a review of discourse analysis and stylistics, Paul Simpson and Geoff Hall (2002) point out that literary discourse needs to be identified 'in terms of multiple intersections among texts, readers, institutions, and sociocultural contexts' (p. 136). In practice, however, while discourse analysis has made a valuable contribution to the study of literature, it has tended to focus on the application of discourse analytical tools to the study of literary texts. The present chapter has taken a rather different approach, drawing on discourse analysis to understand the interpretive literary activity engaged in by readers. Literature itself – the object of the readers' attention – is an archetypally creative form, but I have suggested that readerly activity is also discursively creative in that readers are constructing interpretations, here in collaboration with others, making rather than taking meanings from texts.

I have contrasted a discourse analytic approach focused on analysing interaction between actual readers taking part in a habitual activity, with the imagined reader evident in stylistics and other forms of literary criticism, and experimental readers featuring as subjects in empirical studies. Whereas stylistic analysis tends to provide a singular reading, and experimental research elicits a reading, or contrasting readings, that have been primed by the researcher, the analysis above identifies readings that are rooted in particular interpretive contexts.

A number of points arise from this analysis. For instance, whereas readers do discuss stylistic features (such as characterisation and imagery) as literary accomplishments, they also routinely relate the world of the novel to personal experience and real-world events. The interpretive process is dialogical, with personal experience drawn on to inform interpretation (e.g. considering the actions of a doctor in terms of the power conferred by medical knowledge), but also with literary interpretation leading to reflection on readers' personal lives (the need to picture yourself as a doctor, a wry reflection on the erosion of medical power). In some cases literary reading provides a stimulus for much lengthier discussion of a range of personal, social, political, philosophical etc. issues.

An associated point is that literary reading is closely bound up with reader identities. Reading groups who share an overt identity category – part of the 'who we are' of a group – may frame interpretations in relation to this category (as medics, as women readers, and elsewhere as prisoners, gay men etc.). But a whole range of identities (being a certain age, a librarian, having expertise relevant to a particular novel) may be temporarily invoked in discussion. This isn't to suggest that people read as they do because they are medics, or of a certain age. Features such as a common medical identity may well prime certain readings, but equally, and more creatively, readers draw on their interpretation of a text to make aspects of their identity temporarily, even fleetingly salient, to align themselves with or distance themselves from certain reading positions and to construct themselves in certain ways.

Discourse analysis shows how this plays out, in intricate detail, during the course of an interaction: for instance, how readers cite, elaborate and recontextualise aspects of the text; how readings are sequentially developed and elaborated over a series of utterances; how, during this process, readings may also be challenged or reformulated; how all of this is embedded in interpersonal activity, as readers attend to the management of relations alongside the construction of literary interpretations. This goes some way towards Carter's call for 'enriched understanding of processes of reception' of creative/literary texts. Reading is seen here as both locally contingent and an essentially collaborative accomplishment, far removed from the decontextualised, individual readings elicited in experiments or, for that matter, the popular image of literary reading as an individual activity.

Transcription conventions

In the transcripts above, pseudonyms are used for all speakers.

For ease of reference, speaking turns are numbered sequentially. However, it is recognised that turns here are an artifact of the transcript, and they have no discrete analytical status.

The following transcription conventions are adopted:

(.)	brief pause, untimed
(2.0)	timed pause, duration in seconds
<mmh>	minimal response that does not constitute a separate turn, and where it is not possible to identify the speaker
that's wh-	a dash indicates an incomplete or cut-off utterance
[turned back [and yet he did	square brackets mark the start of overlapping text
(and yet)	curved brackets indicate some uncertainty in transcription
(xxxx)	unclear text that could not be transcribed
((laughter))	double brackets indicate a non-speech element such as laughter, a cough etc.; may also describe the voice quality of the preceding words – e.g. ((laughing voice))
there is no turning back	underlining indicates text that is commented on in the right hand-margin

Notes

[1] While implicit in much literary study, conceptions of the reader are explicitly foregrounded in reader response approaches, focusing on the importance of reader–text interaction (e.g. Fish, 1971; Iser, 1978; Jauss, 1982). However despite this theoretical focus, reader response criticism tends to 'work from the text itself rather than from information about responses' (Culler, 2001 [1981]: 62). This is evident also in more recent approaches to reader response such as cognitive poetics. See further critical discussion in Allington and Swann (2009) and Lillis (2006).

[2] Such textualised conceptions of reading are commonplace in stylistic analysis. A survey of issues of the journal *Language and Literature* over a period of five years (Allington and Swann, 2009) found that readers, or acts of reading, were invoked in 95 per cent of papers presenting an analysis of literary texts. In most cases, however, such references were to an ideal or implied reader.

[3] Short's discussion here concerns a dramatic text (a line from Shakespeare's *Romeo and Juliet*): he does not consider the characteristics of dramatic performances, in which a range of additional contextual factors would need to be taken into account.

[4] Researchers were Joan Swann, Kieran O'Halloran and Daniel Allington. We are grateful to the reading groups who participated in this research and to the AHRC for their support. For information on the project, see http://www.open.ac.uk/dorg/index.shtml.

References

Allington, D. (2007) '"How come most people don't see it?": slashing'. *The Lord of the Rings' Social Semiotics* **17** (1) 45–64.

Allington, D. and Swann, J. (2009) 'Researching literary reading as social practice'. *Language and Literature* **18** (3) 219–30.

Allington, D. and Swann, J. (in press, 2011) 'Reading and social interaction: A critical approach to individual and group reading practices' in *The history of reading, vol. 3: Methods, strategies, tactics,* Crone, R. and Towheed, S. (eds). Basingstoke, Hants: Palgrave Macmillan.

Benwell, B. (2009) '"A pathetic and racist and awful character": Ethnomethodological approaches to the reception of diasporic fiction'. *Language and Literature* **18** (3) 300–15.

Carter, R. (1999) 'Common language: Corpus, creativity and cognition'. *Language and Literature* **8** (3) 195–216.

Carter, R. (2004) *Language and creativity: The art of common talk*. London/New York: Routledge.

Carter, R. (2007) 'Response to special issue of Applied Linguistics devoted to language creativity in everyday contexts'. *Applied Linguistics* **28** (4) 597–608.

Cherland, M. R. (1994) *Private practices: Girls reading fiction and constructing identity*. London: Taylor and Francis.

Coates, J. (1996) *Women talk: Conversation between women friends*. Oxford: Blackwell Publishers.

Cook, G. (2000) *Language play, language learning*. Oxford: Oxford University Press.

Crystal, D. (1998) *Language play*. Harmondsworth: Penguin.

Culler, J. (2001 [1981]) *The pursuit of signs: Semiotics, literature, deconstruction*. Routledge Classics. London/New York: Routledge.

Eriksson, K. and Aronsson, K. (2004) 'Building life world connections during school booktalk'. *Scandinavian Journal of Educational Research* **48** (5) 511–28.

Eriksson Barajas, K. and Aronsson, K. (2009) 'Avid versus struggling readers: Co-construed pupil identities in school booktalk'. *Language and Literature* **18** (3) 281–300.

Fish, S. E. (1971) *Surprised by sin: The reader in paradise lost*. Berkeley/London: University of California Press.

Goddard, A. (2003) '"Is there anybody out there?": Creative language play and literariness in internet relay chat (IRC)' in *Communication research and media science in Europe*, Schorr, A., Campbell, B. and Schenk, M. (eds). Berlin: Mouton De Gruyter.

Gumperz, J. J. (1982) *Discourse strategies*. Cambridge: Cambridge University Press.

Hall, G. (2006) 'Literature as social practice' in *The art of English: Literary creativity*, Goodman, S. and O'Halloran, K. (eds). Basingstoke: Palgrave Macmillan.

Hall, G. (2008) 'Empirical research into the processing of free indirect discourse and the imperative of ecological validity' in *Directions in empirical literary studies*, Zyngier, S., Bortolussi, M., Chesnokova, A. and Auracher, J. (eds). Amsterdam: John Benjamins.

Halsey, K. (2009) '"Folkstylistics" and the history of reading: A discussion of method'. *Language and Literature* **18** (3) 231–46.

Holmes, J. and Stubbe, M. (2003) *Power and politeness in the workplace: A sociolinguistic analysis of talk at work*. London: Longman.

Iser, W. (1978) *The act of reading: A theory of aesthetic response.* Baltimore: Johns Hopkins University Press.

Jauss, H. R. (1982) *Toward an ethic of reception* (trans. T. Bahti). Minneapolis: University of Minnesota Press.

Lillis, T. (2006) 'Readers and writers' in *The art of English: Literary creativity*, Goodman, S. and O'Halloran, K. (eds). Basingstoke: Palgrave Macmillan.

Long, E. (2003) *Book clubs: Women and the uses of reading in everyday life.* Chicago/London: University of Chicago Press.

Maybin, J. and Swann, J. (eds) (2006) *The art of English: Everyday creativity.* Basingstoke: Palgrave Macmillan.

Miall, D. S. and Kuiken, D. (1998) 'The form of reading: Empirical studies of literariness'. *Poetics* **25** 327–41.

Miall, D. S. (2006) *Literary reading: Empirical and theoretical studies.* New York: Peter Lang.

Norrick, N. R. (2000/2001) 'Poetics and Conversation'. *Connotations* **10** (2–3) 243–67.

Radway, J. (1984, 2nd edn 1991) *Reading the romance: Women, patriarchy, and popular literature.* London: Verso Press.

Rose, J. (2002 [2001]) *The intellectual life of the British working classes.* New Haven/London: Yale University Press.

Short, M. (1996) *Exploring the language of poems, plays and prose.* Harlow: Longman.

Simpson, P. and Hall, G. (2002) 'Discourse analysis and stylistics'. *Annual Review of Applied Linguistics* **22** 136–49.

Stockwell, P. (2002) *Cognitive poetics: An introduction.* London/New York: Routledge.

Swann, J. and Allington, D. (2009) 'Reading groups and the language of literary texts: A case study in social reading'. *Language and Literature* **18** (3) 247–64.

Swann, J. (2007) 'Designing "educationally effective" discussion'. *Language and Education* **21** (4) 342–59.

Swann, J. and Maybin, J. (eds) (2007) *Language creativity in everyday contexts.* Special issue of *Applied Linguistics* **28** (4).

Tannen, D. (1984, 2nd edn 2005) *Conversational style: Analyzing talk among friends.* New York: Oxford University Press.

Tannen, D. (1989, 2nd edn 2007) *Talking voices: Repetition, dialogue and imagery in conversational discourse*. Cambridge: Cambridge University Press.

Zubair, S. (2003) 'Women's critical literacies in a Pakistani classroom'. *Changing English* **10** (2) 163–73.

PART TWO

Creativity in professional communication

Traditional discussions of creativity have focused on either works of art or great scientific discoveries, excluding from consideration the realms of business and professional practice. Even the so-called 'creative industries' of advertising, public relations and marketing were in the past deemed by many to involve a type of creativity less worthy of serious scholarly examination. More recently, however, creativity and innovation have become major themes in both the academic study of business and popular discourse about it. Business schools are increasingly preoccupied with how to produce more creative graduates; managers experiment with techniques to bring out the creativity in employees; and governments invest considerable sums in nurturing creativity in key sectors of their nation's economies. Research and popular discourse concerning creativity in business and professional practices have touched upon such topics as product innovation, marketing and customer relations, problem-posing and problem-solving practices in the workplace, inter-professional and inter-organisational communication and convergence, and the challenges of building organisations that can learn and adapt to changing conditions.

Concurrent with this increased interest in business and professional circles in creativity, there has also been an increased interest in discourse studies in the analysis of business and professional communication. Discourse analysts have examined business and professional communication in a wide range of contexts from law to medicine to marketing and from a wide range of perspectives from conversation analysis to genre analysis. This has resulted in a growing appreciation of the complexity and creativity of professional discourse as it operates, as Bhatia reminds us in his chapter, on multiple levels of text, genre, professional practice and professional culture.

The two chapters in this section take up an important theme introduced in the introductory chapter of this volume – the notion that much of creativity in discourse has to do not so much with the formulation of brand new ideas or brand new texts but with the innovative appropriation and re-fashioning of existing texts and the strategic combination of different genres, styles and discursive conventions which can sometimes give rise to 'creative (inter)discourses'. As Bhatia explains it in his chapter, this appropriation of semiotic resources can occur on multiple levels – textual, semantic, socio-pragmatic, generic and professional. Appropriations across texts give rise to *intertextuality* of the kind discussed in the chapter by Jones and his colleagues. Appropriations across professional genres, practices and cultures give rise to the kind of *interdiscursivity* Bhatia describes in his chapter. Along with and related to this attention to appropriation and mixing of texts and discourses in these chapters is an emphasis on the collaborative nature of creativity, the observation that, at least in the contexts described in this section, creativity is centrally a matter of negotiation among different stakeholders in the creative process and often between the demands of different professional cultures.

In the first chapter of this section, Bhatia describes how corporate annual reports combine the discourses of accountancy, law, finance and public relations, and how the interdiscursive relations this combination creates allow companies to 'bend' the conventions of the genre of 'reporting' and present their performance in a positive light. Understanding creativity in professional discourse, Bhatia argues, requires that we consider how multiple discourses, actions and voices interact and are resemiotised in the formation of specific discursive practices within organisations and industries.

In the second chapter, Jones and his colleagues discuss collaborative writing practices in public relations firms, revealing how much of the 'creative content' generated by such concerns is often cobbled together from previously produced texts and company-generated templates. Nevertheless, they show that, far from formulaic, these practices of *bricolage* demand from individuals the ability to successfully integrate old ideas into new paradigms, and from organisations the ability to promote workplace practices and workplace relationships that balance the competing demands of efficiency and originality. In particular, they argue that discourse practices around collaborative writing are creatively used in public relations firms to foster the professional development of individual employees and the creative growth of the organisation.

CHAPTER 5

Creative exploitation of socio-pragmatic space in professional discourse

Vijay K. Bhatia

Creativity in professional discourse is as much a function of the use of lexico-grammatical, rhetorical and discoursal resources as it is of the exploitation of a number of other contextual resources and opportunities that are available to an expert professional writer. The work on creative use of textual resources, such as vocabulary, both technical as well as more general, grammar, and discourse patterns has been paid considerable attention in discourse and genre literature (Fairclough, 1992, 1995; Swales, 1990, 1998; Bhatia, 1993, 1995, 1997, 2000, 2004, to name only a few); however, little attention has been paid to the creative exploitation of contextual and other text-external resources, in particular the innovative use of socio-pragmatic space available to an author. This chapter considers the use of such opportunities by corporate writers in their annual corporate disclosure reports. However, before considering this aspect of creativity in professional discourse, I would like to give a brief account of the nature and function of professional discourse within the context of a genre-based theory of discourse.

Professional discourse seems to operate simultaneously at four somewhat distinct and yet overlapping levels, or concepts of space, identified as discourse as *text*, discourse as *genre*, discourse as *professional practice*, and discourse as *professional culture*. Although these levels in practice may appear to be rather indistinguishable, they seem to be quite distinct from the point of view of genre theory (see Bhatia, 2004, 2008a and 2008b, for a detailed account of this). Different approaches and frameworks for the analysis of discourse have focused on a single or any combination of these levels. In early analyses of discourse, the focus was primarily on texts, and the analytical space within

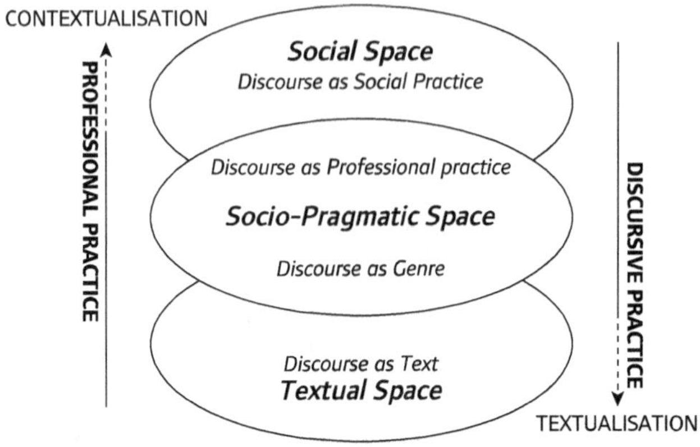

FIGURE 5.1 Multiperspective genre analytical framework
(adapted from Bhatia, 2004)

which observations were made and findings discussed was mostly restricted to textual space. The later approaches however favoured a more generic view of discourse, focusing on genre and action with increasing attention given to textual as well as generic space (Swales, 1990; Bhatia, 1993). Critical discourse analysis combined explorations in textual space with those in social space, thus paying particular attention to specific social institutions, structures and identities. In principle, therefore, it is possible to focus primarily on one or a combination of various levels. Drawing on genre theory (Swales, 1990; Bhatia, 1993), I proposed a three-space model of genre analysis (Bhatia 2004) to account for various levels of professional discourse, as represented in Figure 5.1.

As part of that framework I also proposed a much wider and more interesting concept of space to account for more comprehensive analysis of professional genres, which I termed 'socio-pragmatic space', which incorporates the discursive construction of professional genres as part of specific professional practice to achieve disciplinary as well as professional goals and objectives of a professional community. I would like to further explore 'socio-pragmatic space' in order to study how expert professional writers exploit it in order to bring creativity into their writing of genres, which otherwise may be conventionalised and often standardised in their form, function and social intent (Swales, 1990; Bhatia, 1993). These different levels of analysis of professional discourse within the socio-pragmatic space can be represented visually as in Figure 5.2.

One may notice that, although text and context have been historically assigned varying importance in the analysis of professional genres, the

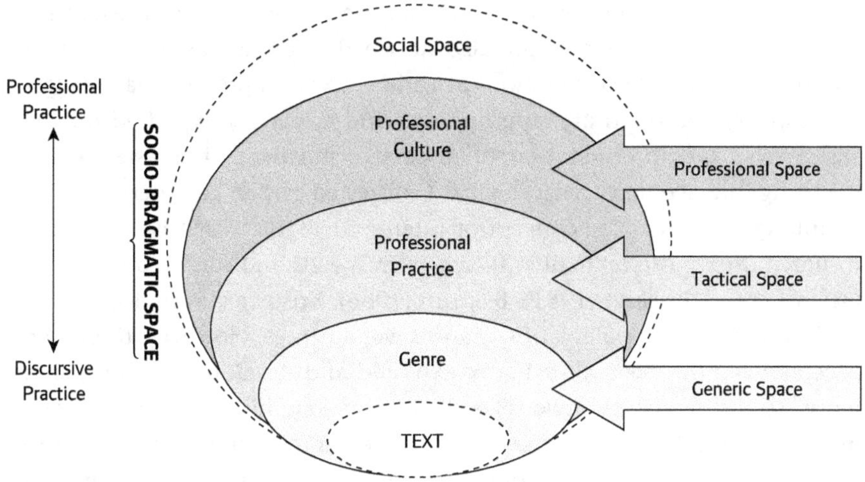

FIGURE 5.2 Diagram 2: Socio-pragmatic space in professional discourse
(adapted from Bhatia, 2004)

focus, in more recent analyses of professional genre (Swales, 1998; Bhatia, 2004, 2008a, 2008b, 2010) has been shifting towards increasing attention to context, thus redefining genre as a configuration of text-external and text-internal factors, highlighting two kinds of relationships involving texts and contexts. In my view, the relationship between and across texts, focusing primarily on text-internal properties, is intertextual in nature, whereas interactions between and across text-external resources, such as genres, professional practices, or professional cultures, should be seen as interdiscursive in nature (Bhatia, 2010). Intertextuality has been given quite a bit of consideration in discourse and genre theory; interdiscursivity, however, has attracted relatively little attention, especially in genre theory. Drawing evidence from the analysis of corporate disclosure practices, I would like to focus on how interdiscursivity in professional genre is creatively exploited by expert corporate players to meet not only their professional goals but also their 'private intentions' (Bhatia, 2004).

In another sense, this above-mentioned realisation of professional discourse also highlights two other kinds of relationship helpful for analysis, one representing the text-context relationship, and the other representing the discursive-professional practice relationship. Interestingly, these two relationships seem to be the function of text-internal and text-external semiotic resources and constraints. As briefly alluded to earlier in the chapter, text-external resources include the conventions that constrain generic constructs as well as professional practices, and perhaps more appropriately specific disciplinary cultures that motivate these practices, both discursive as well as professional. We may also notice that textual as well as other semiotic resources and conventions at various levels of professional engagement can be, and often are, appropriated and creatively exploited for the construction and interpretation of discursive practices, thus establishing interesting interactive patterns of what I suggested earlier, i.e. intertextuality and interdiscursivity. The concept of interdiscursivity, which is sometimes subsumed under intertextuality, is not entirely new and can be traced back to the works of Foucault (1981), Bakhtin (1986), Kristeva (1980), Fairclough (1995), Candlin and Maley (1997), and several others. However, these two concepts have not been sufficiently explored and developed within genre theory. To distinguish the two related concepts, one may say that, whereas intertextuality refers to the use of prior texts transforming the past into the present often in relatively conventionalised and somewhat standardised ways, interdiscursivity, on the other hand, refers to more innovative attempts to create various forms of hybrid and often relatively novel constructs by appropriating or exploiting established conventions or resources associated with other genres and practices. Interdiscursivity, in my view, accounts for a variety of discursive processes and professional practices, often resulting in 'mixing', 'embedding', and 'bending' of generic norms in professional contexts (see Bhatia, 1995, 1997, 2000, 2004 for a more detailed account). More importantly for our discussion here, interdiscursivity can be viewed as appropriation of semiotic resources (which may include textual, semantic, socio-pragmatic, generic and professional) across any two or more of these different levels, especially those of genre, professional practice and professional culture. Appropriations across texts thus give rise to intertextual relations, whereas appropriations across professional genres, practices and cultures constitute interdiscursive relations.

In order to develop a comprehensive and evidence-based awareness of the motives and intentions of such disciplinary and professional practices (Swales, 1998), one needs to look closely at the multiple discourses, actions and voices that play a significant role in the formation of specific discursive practices within relevant institutional and organisational frameworks, in addition to the conventional systems of genres (Bazerman, 1994), often used to fulfil the professional objectives of a specific professional community.

After this brief account of the creative manipulation of corporate discourse, I would now like to focus on a specific instance of interdiscursive exploitation of tactical space to realise a desired effect and interpretation of the corporate annual report, especially keeping in mind the private intentions of the corporation.

Corporate annual reports[1]

This chapter is based on the work carried out under a project focusing on 15 Hong Kong Stock Exchange listed companies, in which an attempt was made to critically examine corporate disclosure practices through corporate disclosure genres, variously reported in my earlier works (Bhatia, 2008a, 2008b, and 2010). The focus in this chapter is not on what the companies did or did not disclose in a particular year, but, more importantly, on the hidden intentions and motivations for such practices, thus highlighting interesting and creative use of socio-pragmatic space generally available within the broader interpretation of corporate annual or periodical reports. A critical aspect of this study is to focus on the creative use of text-external resources that make it possible for corporate writers to persuade and convince a varied audience of stakeholders, in particular minority shareholders, to accept their accounts, estimates, and projections of the future performance of the corporation in question in the prevailing competitive corporate culture of sustaining shareholders' interest and investment in the corporation.

The companies were categorised into three main groups based on their performance during the period 1998 to 2005. They were categorised on the basis of their good, moderate or poor performance. The selection of the corpus, in particular the performance of companies as revealed in corporate periodic and annual reports and several other documents, was made in consultation with specialists in business and finance, keeping in mind their overall turnover, progress during the specified period, and return on equity, in addition to various other indicators.

The main corpus for this multidimensional as well as multiperspective genre analysis (Bhatia, 2004) consisted of corporate disclosure documents, which included annual and other periodical performance reports, press releases by these companies about their performance, and other communications from management to (minority) shareholders. These comprised sections of annual and other interim reports, press releases by these companies, earnings/results announcements, shareholder circulars, notes from general meetings, transcripts/slides from speeches/presentations, and a number of other corporate publications such as the company newsletters.

In addition to this primary corpus, there was an additional corpus of secondary data, which included newspaper reports on corporate performance, journal articles highlighting or reviewing the performance of these companies, assessments of the performance of these companies by rating agencies, such as Standard and Poor's, and other financial institutions, such as the Monetary Authority of Hong Kong, etc. The main function of the secondary data, which is primarily drawn from newspapers, magazine articles, and internet websites of rating agencies and institutions, incorporating assessments, reviews, and evaluations of performance by outside rating agencies, was to provide additional informed third-party perspectives on the performance of these companies. Besides this secondary corpus, the project also sought perceptions of participants and specialists in corporate disclosure practices, which included not only accountants and public relations experts, but also expert members of the rating agencies as well as experienced minority shareholders through semi-structured interviews. However, in this chapter I would like to consider the form, function and value attached to only one kind of corporate genre, that is, the annual report.

Corporate annual reports have long been considered the pulse of corporate realities. Their main purpose is to inform their shareholders about the performance and health of the company, specifically its successes and failures, current problems, and prospects for its future development. However, corporate annual reports and other disclosure documents seem to be changing in their function from 'informing and reporting' to increasingly 'promoting' the companies by a strategic underplaying of corporate weaknesses, often 'bending' the norms of corporate disclosure genres.

One of the interesting findings of the analysis undertaken under the project is that the corporate annual report is a typical combination of at least four interesting but different discourses included in the same document, all of which, partly because of the interdiscursive relations between them and partly through the creative use of specific lexico-syntactic as well as socio-pragmatic resources, are cleverly exploited to 'bend' the norms and conventions of 'reporting' to promote a positive image of the company, even in adverse and challenging economic circumstances. The companies through these annual reports often seem to negotiate the tension between the need to underplay a relatively weak corporate performance and to project the expectations of good performance (speculation about the future outlook), especially in contexts when uncertainties about future economic and corporate growth and performance threaten encouraging prospects in the coming years. In addition, they also need to conform to the legal requirements of disclosure of information for the benefit of corporate stakeholders. So essentially the corporate players need to handle four different kinds of discourses within the same genre: this is conventionally known as a corporate annual report. The four kinds of discourses are:

- *Accounting discourse*, which forms a major part of the Annual Reports, duly endorsed, certified by public accountants.

- *Discourse of finance*, in the form of what is conventionally known as the financial review section of the report.
- *Public relations discourse*, in the form of the chairman's letter to shareholders, for which public accounting firms do not take any responsibility.
- *Legal discourse*, which forms a major part of disclaimers, often necessary to comprehend the full implications of the information disclosed in the report.

Obviously, the most important section of the annual report consists of the accounting information, which is mostly nothing more than the numbers displayed in the form of tables, graphs, and calculations of various kinds. These numbers, though originally proposed by the corporation, are certified by public accountants to be true and honest representations of the profits and losses of the corporation in question.

The discourse of economics and finance, on the other hand, offers a review of the company's performance based on the numbers certified by the public accountants; however, this section is not certified by public accountants. This section is written by specialists in financial management in the corporation, who generally offer the corporate view of the performance of the company, apparently on the basis of the information included in the accounting section, which they often refer to. However, there may not always be any explicit relationship between the two forms of discourse. They are positioned in the same interdiscursive space, perhaps to give the impression that one is a true representation or at least a reflection of the other, rather than an authoritative interpretation of the other. This strategic appropriation or manipulation of interdiscursive space creates an impression in the mind of uninitiated readers, particularly the minority shareholders, who really do not understand the complexity of numbers in the accounting discourse, that the one they understand better is a seemingly reliable interpretation of the one that is beyond their capacity to understand. If we take one more step beyond these two business discourses and consider the third form of discourse, which most readers can understand, we find the Chairman's statement, which, in reality, is nothing more than public relations discourse, giving the predictions and projections based on the company performance in the reporting year. Once again, it can be considered at least twice removed from the reality of the accounting discourse, but generally believed to be based on the certified numbers given elsewhere in the annual report. It is interesting to note that the accounting discourse, which is accessible only to insiders, is strategically appropriated to construct the discourse of finance (sometimes called 'management's analysis and discussion'), which is relatively more accessible than the former, and both of these are further appropriated to construct the discourse of public relations (Chairman's Letter or Statement), which is easily accessible to insiders as well as to outsiders. Let me illustrate this relationship by considering examples of such discourses.

Accounting discourse

> We have audited the consolidated financial statements of XXX set out on pages XX to XX, which comprise the consolidated and the XXX's statements of financial position as at Date . . . , and the consolidated income statement, consolidated statement of comprehensive income, consolidated statement of changes in equity and consolidated cash flow statement for the year then ended, and a summary of significant accounting policies and other explanatory notes . . .
>
> In our opinion, the consolidated financial statements give a true and fair view of the state of the affairs of the Bank and the group as at 31 December 2009 and of the profit and cash flows of the group for the year then ended in accordance with Hong Kong Financial Reporting Standards and have been properly prepared in accordance with the Hong Kong Companies Ordinance
>
> (XXX's Annual Report 2009)

These extracts constitute much of the textual material we find as part of the accounting discourse. These paragraphs contain preliminary information and not what is often associated with substantive information on corporate disclosures. The rest of the pages contain statistical information consisting of financial statements, originally prepared by the company's financial department and certified by the Public Accountants. The numbers are mostly in tabular forms and displays of various other kinds, a section of which is shown in Table 5.1.

In comparison with this typical accounting information, let me take up an example of the second kind of discourse, which I have referred to as the discourse of finance, often given various names in different corporate disclosure annual reports.

TABLE 5.1 Financial reporting

	2009	*2008*
For the year	HK$m	HK$m
Net operating income before loan impairment charges	xxx	xxx
Profit before tax	xxx	xxx
Profit attributable to shareholders	xxx	xxx
At the year-end		
Shareholders' equity	xxx	xxx
Total equity	xxx	xxx
Total capital base	xxx	xxx
Customer accounts	xxx	xxx
Total assets	xxx	xxx

Discourse of finance (financial review)

XXX reported profit before tax of HK$ xxx million, a decrease of x.x per cent over 2008 primarily due to the fall in net interest income which was partly offset by increased trading income and lower loan impairment charges.

Net interest income decreased by HK$ xxx million, or xx.x per cent, compared with 2008 as falling interest rates in the second half of 2008 impacted all business lines throughout 2009. In Personal Financial Services and Commercial Banking the relatively lower funding cost of the asset book and the increase in customer deposits of xx.x per cent was more than offset by the year-on-year narrowing of interest rate spreads which impacted the overall margin. In Personal Financial Services the mortgage book continued to perform strongly . . . The average loan-to-value ratio was xx.x per cent in respect of new business. In Commercial Banking customer loans and advances fell by x.x per cent as market demand reduced and trade fell. However, there was an increase in trade volumes and lending activities in the second half of the year. Despite the challenging environment, XXX continued to support local businesses . . .

(XXX's Annual Report xxxx)

As we can see, this discourse is different from the former in that it is more of a discussion and analysis of facts and figures, which form a substantial part of the accounting discourse. Since the two are put together and there are frequent references to accounting information, the readers are likely to get the impression that the discourse of finance must be a true interpretation of the accounting discourse. However, it is likely that any such resemiotisation (Iedema, 2003) of the accounting information may not necessarily be a consistent and true representation of the statistical information and may even lead to varying interpretations, but the reader is less likely to question it, if the two sections share the same socio-pragmatic space. In this case, therefore, interdiscursivity across the two rather different discourses comes from the exploitation of the same socio-pragmatic space. The financial review thus is a resemiotisation of a 'convenient' selection of the accounting discourse meant for those corporate stakeholders who do not have enough either expertise or time to understand the full implications of the accounting discourse. There is an expectation that this section of economic discourse will take into account the main features of the annual report, and there is also an expectation that there will be an adequate degree of referential help available for the reader to go through relevant sections of related textual and numerical information.

Let us make the picture a bit more complex by bringing in yet another form of discourse from the annual report genre, that is, the discourse of public relations.

Public relations discourse

> XXX has remained profitable, financially strong and independently owned by our shareholders. It is testimony to the quality and strength of XXX's management team . . . On an underlying basis, and excluding the impact of the goodwill impairment recorded in 2008, pre-tax profit was US$ xxx billion, xx per cent higher. On a reported basis, profit before tax was US$ xxx billion, down xx per cent, in part due to the reversal of fair value accounting gains on our own debt.
>
> That XXX has reported a pre-tax profit in all three years since the onset of the crisis should be a source of great confidence to our shareholders, our depositors and all of our customers. Our track record of delivering results through adversity, and at all stages of the economic cycle, remains intact. We continued to enhance our financial strength during 2009. We strengthened our capital base by US$ xxx billion through underlying profit generation. . . . Throughout the crisis, our strategy has remained clear to build on our position as the leading international and emerging markets bank. We have also never forgotten

> that it is our responsibility to make a real contribution to economic and social development, and that our ability to do so is fundamental to our success in delivering sustainable value to our shareholders.
>
> (XXX's Annual Report xxxx)

This discourse, of course, is a further resemiotisation of the earlier discourses, and is meant to reassure stakeholders that the performance of the company is reasonably strong and the future seems even better than the past. There are a few references to facts and figures, which form an integral part of the two earlier discourses, and most of the estimations and predictions are based on impressions and hopes of the chairman.

The three discourses, which are taken from the same annual report of a public listed company, are placed in a particular order within the socio-pragmatic space of the same corporate document, co-constructing the intended meaning and serving a rather typical corporate objective of informing stakeholders as well as public monitoring authorities about the performance of the company in the preceding year while at the same time giving a rather positive impression of the company's performance to its shareholders in order not to disturb the 'sentiment of the investors' (according to a specialist informant interviewed by the project investigator) or to precipitate an undesirable downward trend in share price movement. It is interesting to note that the three discourses are rather different in accessibility to the report's intended readership, particularly to outsiders like the minority shareholders as against the insiders, i.e. specialists such as institutional investors, financial experts and business managers. The three discourses also vary considerably in terms of the technicality and complexity of information, but the most interesting aspect of such reports is that it is the technically complex accounting discourse that has the maximum amount of credibility, in that it is this discourse which is certified by the public accounting firms. As the discourses become less technical, they become less credible as they are based on the interpretations of the certified accounting discourse. As a senior public accountant (interviewed by the researcher) points out:

> The accountant's job is to note any inconsistency in the company reports and . . . to go through the banking mechanism of the management . . . Accountants are not responsible for the forecasting of the company's future.

So the certification by a public accounting firm lends a strong measure of credibility to corporate disclosures, whereas the public relations discourse is a typical public relations exercise, which takes advantage of the other discourses strategically placed within the same socio-pragmatic space, or more centrally tactical space. A senior public relations manager interviewed about her involvement in the chairman's letter in the annual report candidly pointed out:

> . . . the Chairman's Statement in a company report (is) the PR's job . . . At times when profit is not so good, her role as a PR manager is to manage the sentiments of the public and the shareholders so as not to make any dramatic share price movement. This involves management of tone and manner of writing the reports as well as meeting the expectations of the Management.

The three discourses thus tend to serve two very different purposes: the accounting discourse tends to report accurately and factually on the basis of financial evidence about the past corporate performance, whereas the chairman's letter is meant to promote a positive image of the company to its shareholders and other stakeholders in order to sustain their confidence in future corporate performance. Both these discourses are products of two very different corporate practices: one is centrally located in the conventional and legally required practice of auditing corporate results, and the other is an instance of marketing and public relations practice. They also use very different textual resources and rhetorical strategies: one uses numerical data consistent with audit and accountancy practices, whereas the other makes use of promotional as well as forecasting rhetorical strategies. However, it is interesting to note that the two discourses are strategically placed in the same document, that is, the annual report, thus establishing an interesting interdiscursive relationship between these two different corporate genres. The real motivation for placing the two discourses within the boundaries of the same corporate annual report is that such textual proximity is likely to lend marketing and public relations discourse the same factual reliability and hence credibility that is often presupposed based on the use of numerical data. The public relations discourse, on its own, is likely to be viewed by the intended audience of (minority) shareholders as a promotional effort, but when it is placed in the discoursal context of the accounting discourse, which is often viewed as more evidence-based, factual and therefore reliable, it is likely to raise the legitimate presupposition that it may be drawing its conclusions from the accounting numbers that are certified by a public authority accepted by the controlling government agencies. Many of the minority shareholders, whose numbers have increased considerably in recent years, often lack the expertise and even the linguistic skills to fully understand the implications of the accounting discourse in the annual report, but when they see the chairman's letter to shareholders in the same report, which is 'assumed' to be based on the accounting data, they are likely to take at least some of the predictions and speculative statements in the letter rather more seriously than otherwise.

Since the public relations (i.e. promotional) discourse signed by the chairman of the company projecting a positive image of the company is placed in the context of the other two discourses, one assumes that this one is also truly based on the other two, which may not always be the case. However, this strategic exploitation of the socio-pragmatic space within the genre of annual corporate reports somehow tends to lend further support to the estimations, predictions and conclusions presented in the report. Nevertheless, Stock Exchanges and other public monitoring bodies everywhere have listing rules which contain specific provisions to protect minority shareholders so that they are given sufficient information from proper ongoing disclosures to make informed decisions on their investments. Baker and McKenzie, in their April 2002 'Corporate Alert', cite an interesting case from Australia, *GPG (Australia Trading) Pty Ltd v GIO Australia Holdings Ltd*, where a shareholder successfully sued GIO for losses for publishing a misleading or deceptive notice in relation to corporate disclosures. Corporations, on their part, guard against any possible (intended or unintended) overestimations, unsupported predictions, or speculative statements in their PR discourses, and even elsewhere in the financial management discourse sections, so that the annual report often includes yet another form of discourse that is popularly known as legal disclaimer, of which the following is a typical example.

Legal discourse (disclaimer)

> The Annual Report and Accounts 2009 contains certain forward-looking statements with respect to the financial condition, results of operations and business of HSBC. Statements that are not historical facts, including statements about HSBC's beliefs and expectations, are forward-looking statements. Words such as 'expects', 'anticipates',

> 'intends', 'plans', 'believes', 'seeks', 'estimates', 'potential' and 'reasonably possible', variations of these words and similar expressions are intended to identify forward-looking statements. These statements are based on current plans, estimates and projections, and therefore undue reliance should not be placed on them. Forward-looking statements speak only as of the date they are made, and it should not be assumed that they have been revised or updated in the light of new information or future events . . .

Thus whatever is claimed in the corporate disclosure documents is systematically disclaimed through typical legal statements of this kind. If readers, especially minority shareholders, are unaware of the content of such disclaimers and tend to read the predictions, projections and other speculative claims made through the annual report without disclosing the detailed investment risks, they can be and often are shocked by the corporate results in succeeding years.

This rather brief account of corporate disclosure practices clearly underpins the importance of corporate practices in the study of corporate genres. A purely textual or discoursal study of lexico-grammatical, semantic-pragmatic, and rhetorical or discourse organisation is unlikely to reveal the intricacies of the construction, interpretation, use and especially exploitation of these corporate genres in achieving their corporate objectives within the requirements of disclosure practices imposed by corporate governance bodies and rating agencies. In order to fully appreciate and understand the complete implications of such discursive practices, in particular to investigate how these corporate players creatively 'bend' generic norms and conventions to communicate modest or weak corporate performance without undermining investor confidence in their company, one needs to combine the analysis of professional discourses with the disciplinary practices of the professions in question. The data analysed implies that, although it did not directly involve appropriation of generic resources across two genres, by placing public relations discourse within the socio-pragmatic space of the accounting discourse the readers are encouraged to draw positive and hopefully reliable implications from this interdiscursive placement of four different kinds of discourse within the same socio-pragmatic space.

The study also indicates that it is possible to understand a specific professional practice, e.g. how corporate leaders strategically and creatively meet the statutory requirements of periodical disclosures of corporate performance without undermining investor confidence at difficult times, by analysing the discourses in and of corporate disclosure practices, as these disclosure practices are constituted by a relevant set of professional discourses and genres tactically exploiting, through resemiotisation, the available socio-pragmatic space.

Conclusion

In this chapter I have made an attempt to explore the nature of creative exploitation of interdiscursive socio-pragmatic space, claiming that it is central to the understanding of professional genres and practices, thus arguing for a move towards a more critical genre analysis (see also Bhatia, 2008a, 2008b, 2010). Focusing on genre-based professional, corporate and institutional actions, this chapter has also made an attempt to examine the interrelationship between discursive practices (constructing, interpreting and using professional genres) and professional practices (managing professional activities such as corporate disclosures, public relations, negotiating investor confidence, selling corporate performance, etc.) in typical professional and corporate contexts. At a more theoretical level, I have made an attempt to underpin the importance of interdiscursivity and resemiotisation in critical genre analysis, highlighting the notion of tension between 'generic integrity' (Bhatia, 1993, 1995, 2004) and 'appropriation of generic resources' (Bhatia, 1997, 2004) in professional discourse. The study argues for critical genre analysis of discursive and professional actions within professional cultures by focusing on 'interdiscursivity' as interaction between discursive and professional practices in the context of specific professional, corporate and institutional cultures, which may be viewed as a move towards critical genre analysis.

Notes

[1] Research to be reported in this paper was supported by the RGC CERG grant from The Government of HKSAR [Project No. 9041056 (CityU 1454/05H)] on *Genre-bending in corporate disclosure documents.*

[2] The illustrative texts in this chapter, though part of publicly available documents, have been sanitised to preserve confidentiality. 'XXX' refers to the name of the company, whereas 'xxx' refers to other information, generally numerical, which has been deleted.

References

Bakhtin, M. M. (1986) *Speech genres and other late essays.* Austin: University of Texas Press.

Bazerman, C. (1994) 'Systems of genres and the enhancement of social intentions' in *Genre and the new rhetoric,* Freedman, A. and Medway, P. (eds). London: Taylor and Francis, 79–104.

Bhatia, V. K. (1993) *Analysing genre – language use in professional settings.* London: Longman.

Bhatia, V. K. (1995) 'Genre-mixing in professional communication: The case of "private intentions" v. "socially recognised purposes"', in *Explorations in English for professional communication,* Bruthiaux, P., Boswood, T. and Bertha, B. (eds). City University of Hong Kong, Hong Kong, 1–19.

Bhatia, V. K. (1997) 'Genre-mixing in academic introductions'. *English for Specific Purposes* **16** (3) 181–96.

Bhatia, V. K. (2000) 'Discourse of philanthropic fundraising'. *New directions in philanthropic fundraising* **22** 95–110.

Bhatia, V. K. (2004) *Worlds of written discourse: A genre-based view.* London: Continuum International.

Bhatia, V. K. (2008a) 'Genre analysis, ESP and professional practice'. *English for Specific Purposes* **27** 161–74.

Bhatia, V. K. (2008b) 'Towards critical genre analysis' in *Advances in discourse studies,* Bhatia, V. K., Flowerdew, J. and Jones, R. (eds). Routledge, 166–77.

Bhatia, V. K. (2010) 'Interdiscursivity in Professional Communication', *Discourse and Communication,* **4** (1), 32–50.

Candlin, C. N. and Maley, Y. (1997) 'Intertextuality and interdiscursivity in the discourse of alternative dispute resolution' in *The construction of professional discourse,* Gunnarsson, B-L., Linnel, P. and Nordberg, B. (eds). London: Longman, 201–22.

Fairclough, N. (1992) *Language and power.* London: Longman.

Fairclough, N. (1995) *Critical discourse analysis: The critical study of language.* London: Longman.

Foucault, M. (1981) *The archaeology of knowledge.* New York: Pantheon Books.

Iedema, R. (2003) 'Multimodality, resemiotization: extending the analysis of discourse as multi-semiotic practice'. *Visual Communication* **2** (1) 29–57.

Kristeva J. (1980) 'Word, dialogue and novel' in *Desire in Language*, Kristeva J. (ed.). Oxford: Blackwell, 64–91.

Swales, J. M. (1990) *Genre analysis: English in academic and research settings*. Cambridge: Cambridge University Press.

Swales, J. M. (1998) *Other floors, other voices: A textography of a small university building*. London: Lawrence Erlbaum Associates, Publishers.

CHAPTER 6

Creative collaboration in the public relations industry

Rodney H. Jones, Vijay K. Bhatia, Stephen Bremner and Anne Peirson-Smith

This chapter explores creativity in the collaborative writing practices of public relations firms. In the previous chapter in this section, the focus was on the hybrid, intertextual and negotiated character of creative texts in corporate and professional contexts. While this chapter will also be concerned with hybridity and intertextuality in creativity, its main focus will be less on creative products and more on the *social processes* which result in this creative hybridity, on the kinds of discursive tools people use to support these processes, on the kinds of social relationships that facilitate them, and on the ways novices in professional contexts are socialised into them.

Creativity in the public relations industry is rather different from literary creativity of the kind discussed in the first section of this book, not just because of its practical, economic orientation, but also because, rather than being seen as a property of specific texts, it is more likely to be associated with longer-scale processes which involve a large number of interrelated texts. Creativity is more a matter of how these multiple texts are coordinated and work together to accomplish a particular set of goals within campaigns that traditionally involve multiple creative activities including generating ideas, conducting research, composing press releases, pitching media outlets, writing speeches, arranging appearances on radio and television news and talk shows, developing case studies, and assembling press kits (Aronson *et al.*, 2007). Individual texts are not necessarily seen as creative in themselves, but deemed creative insofar as they contribute to the overall objectives of the campaign. In other words, texts are seen as parts of 'coordinated (and) purposeful actions' designed to solve particular problems and meet particular overall objectives (Reddi, 2009: 400). Creativity is located not so much in discursive products (texts) as in the coordinated *discursive practices* through which different kinds of texts are crafted and deployed in different ways over the course of a campaign.

The theoretical framework through which we will approach our examination of these linked coordinated actions comes from mediated discourse analysis (Norris and Jones, 2005; Scollon and Scollon, 2004), an approach which sees all social actions as taking place at a *nexus* of three interrelated conditions: the first is the *discourses in place,* the kinds of texts and other cultural tools available to social actors in a particular setting at a particular time with which to take action; the second is the *interaction order,* the relationships among the different social actors involved in the action; and the third is the *historical bodies* of the actors themselves, their knowledge, experiences and mastery of particular social practices. How any social action turns out is ultimately a matter of how these three elements interact.

In the context of public relations practice, these three elements are crucially important when examining creative processes. Such processes, for example, almost always involve the appropriation and adaptation of previously produced texts – *discourses in place* – in the service of new situations and communicative goals. Creative processes in public relations firms are also usually collaborative affairs, taking place within complex *interaction orders* in which individuals must manage multiple face systems ranging from the rigidly hierarchical to the more egalitarian. Finally, creativity depends at the end of the day on the ability of individuals to think of and implement good ideas, an ability which comes from previous training and experience involving both formal education and informal socialisation into the creative practices and 'culture' of the firm for which they are working. This combination of mastery, knowledge and experience which come together within particular social actors is what mediated discourse analysis refers to as the *historical body*.

The advantage of mediated discourse analysis as a theoretical perspective for our purposes here is that it provides a way to examine how these three elements work together and affect one another at the level of concrete social actions within the workplace: how *discourses in place,* for example, help to support different kinds of social relationships and help to socialise individuals into institutional practices of creativity; how *interaction orders* help to determine the ways *discourses in place* are used and who can use them, and how they either encourage or constrain creative contributions by individuals; and how the experiences of individuals and the degree to which they have been socialised into the culture of the firm shape the ways in which they are able to use the texts and tools available to them and how they are able to manage their relationships with those with whom they are collaborating.

The data for this discussion come from a two-year-long ethnographic study of three public relations firms in Hong Kong, a small, local operation, a medium-sized regional firm, and a large multinational concern. In the course of the study, a research assistant was placed as an intern in each firm for a period of three months, during which time she conducted participant observation, collected texts related to the campaigns in which she was involved, tape-recorded various formal and informal interactions in the workplace, and kept a detailed reflective diary about her experiences and her ongoing socialisation into the corporate cultures of the companies in which she was placed. This ethnographic data was supplemented by more formal interviews with various staff from senior managers to novice copywriters and consultants at the companies where the participant observation took place as well as a number of other firms[1].

In what follows we will draw from these data to describe the ways discourses in place, interaction orders and historical bodies interact in the management of PR campaigns, and how different configurations of these elements can either facilitate or constrain creativity.

Creativity in the public relations industry

Our interest in creative processes is more than merely theoretical, but rather grows out of the preoccupations of the people we were studying, public relations practitioners, whose main concerns centred around rather practical issues: how they could maximise profits, solve concrete problems for their clients, manage their staff effectively and measure the outcomes of their efforts? The managing director of the regional PR firm we studied put it this way:

> My main concern is how can we get the highest multiple on the salary that we pay those people to the overall revenue that we bring in. No matter what other people might tell you, it's really hard to make more than about 10 per cent net after paying taxes . . . and everything else . . . So, there isn't a very high margin for error. And so when it comes to creativity, we have to think about how we can deliver that in a relatively *un*-time-consuming way . . . We may have the most creative idea but it might not be economical . . . How can we have a system and structure that gives us the right level of client service for the amount of money they are paying? . . . How do we get to that good idea quickly?

Another interviewee, an account executive in a multinational firm, said this:

> Everybody emphasises the need and importance of creativity, but everybody emphasises again, creativity, for the sake of it, is no use and no good definitely . . . you have to make sure that creativity can ultimately deliver results and solve business problems . . . creativity has to be, um, practical, has to be focused. It has to be focused, and to make sure that it delivers business results, it has to be measurable . . . How can it be measured? How can we be accountable for the investment that the client puts in to that type of creativity?

The common theme of these two remarks and many others from our participants was that creativity was largely a matter of efficient problem solving. Whether or not an idea or strategy was deemed creative was not just a function of the idea's novelty or merit, but also of its practicality and cost-effectiveness. Related to this belief was the widespread feeling that creativity as it is often defined in other domains – as originality or innovation – was not only overrated but also not always valued by clients themselves. As one manager of a large regional firm said, 'Agencies think (clients) want creativity, but actually mostly clients don't. They want no surprises, good budgeting and an account team that understands the business and that doesn't change very often'.

Because of both the economic imperatives surrounding their creative work, and a certain scepticism regarding creativity in general, senior management in the firms we studied were most interested in how we might be able to help them to answer three interrelated questions: 'How can creativity be efficiently supported in the organisation?', 'How can we get people to work together to be creative?' and 'How can newer employees be effectively and efficiently socialised into creative practices within the organisation?' Put into the parlance of mediated discourse analysis, these questions might be rephrased as: 'What kinds of *discourses in place* (creative resources) need to be available?', 'What kinds of *interaction orders* support effective collaboration at different stages of the process?', and 'How do the creative practices in a firm come to be submerged into the *historical bodies* of employees?'

Discourses in place

Bakhtin's (1991) famous proclamation that all texts are heteroglossic, that they are composed through the combination of multiple previous utterances, is perhaps nowhere more evident than in the public relations industry. In all of the firms we studied, the creation of texts, whether they were proposals, press releases, speeches or brochures, was often more a matter of inventive *bricolage* than of originality. As one account executive in the medium-sized firm we studied put it:

> If some text can be reused, then we reuse it. We may just change some of the information like the name or background of the clients, because what the clients want is sometimes very similar. For proposals, we may only have to change the services information and the budget. If it's a press release, we sometimes just need to update the facts and figures.

Under these circumstances, the quality and efficiency of creative processes depend crucially on the quality and availability of previously thought-of ideas, previously articulated utterances and previously written texts – in other words, of *discourses in place*.

In the firms we studied, these *discourses in place* generally came in two forms: 'templates' and 'boilerplate'. Templates consist of prefabricated frameworks for texts in which new information is inserted during the composition process. Boilerplate consists of paragraphs or longer texts, which can be inserted into new texts with little or no alteration. Most writing tasks we observed involved using *both* templates, previously designed frameworks, and boilerplate, previously written texts, at the same time.

Normally, however, the practitioners that we spoke to made a rather strong distinction between templates and boilerplate – frameworks and content – and this distinction often had to do with their notions of creativity. It was generally accepted that nearly every text would have a set template, that is, a set structure with certain ordered communicative 'moves' (Bhatia, 1993) which usually could not be deviated from. In fact, often the first step in the composing process involved selecting the appropriate template for the document, and firms often made a point of training employees in using these templates. Content, on the other hand, was seen as much more variable. Whereas some boilerplate could be reused quite easily, other boilerplate had to be redrafted to fit appropriately into its new context. Ultimately, success in this form of intertextuality depends on striking a balance – matching the right template to the right boilerplate. As one account executive put it, 'Sometimes you cut and paste but actually it's completely wrong. Because the framework is ok, but the content is completely wrong, or the content is good, but it's in the wrong framework'.

There were two main sources for such *discourses in place*: the firm itself and the clients it served. Clients were much more likely to provide boilerplate. One consultant for a small public relations firm, for example, described writing press releases this way:

> Normally you do not have to draft the press release because the client has already sent you the information. What you need to do is just to modify the information a bit and present it in a press release format.

In this case, the 'press release format' normally used by this firm acted as the template within which the boilerplate provided by the client was inserted. Templates, on the other hand, almost always come from the firm and constitute an important part of the firm's discursive culture, heritage and branding. The purpose of templates is not just to assist the firm in producing texts more efficiently, but also to assist readers, especially media professionals, in using these texts more effectively as genres. As one senior account executive put it:

> Our templates are like our company brand. When people get stuff from us, they know instantly what to do with it. There's some information, for instance, that we always put at the beginning or in particular parts of the text. People who get used to receiving our press releases will know instantly where to find what information.

Some companies make templates and boilerplate, in the form of previously authored press releases, proposals and other texts, available to employees online through secure servers or on shared hard drives. A senior manger of the large multinational firm described his company's practices like this:

> So for example if you're trying to put together a new business proposal, basically you just log on and you can download every case study that (name of firm) has ever done on that kind of product or service and just dump the information into your own presentation without having to ask somebody or put out a mass email, and that's being constantly updated with every single thing that comes out of this firm, and we have dedicated people who're actually managing these resources to make sure they're being built up and that the right resources are easy to find.

Creating new texts, then, was seen as both a *retrospective* and an *anticipatory* process, a process of looking back at previously authored texts and appropriating material from them, and of looking forward to a time when the current text might itself be appropriated for some future campaign. Creativity was thus seen not so much as a matter of creating new texts as it was a cyclical process of drawing on and replenishing the firm's *archive* of texts. 'As business people', one senior manager said, 'we want to reuse things, you know, we have a really good idea or concept we want to turn that into something that we can use again and again and again'.

Nearly all of the practitioners we talked to, however, cautioned against assuming that the appropriation of previously written texts or previously designed templates was either easy or uncreative. It involves, they told us, complex practices of decision-making which include not just selecting the best texts to appropriate but also adapting them in creative ways to fit new products or new goals. The same senior manager quoted above, for example, said:

> When you're writing something, you should never have to start from scratch, so you have to start by thinking, right, what have we done before that's applicable here? If the objective is to launch a new car in China you might actually find that the proposal that we did to launch a new plane in China is the most appropriate template, or if it is to launch a regional line of mobile phones, we actually launched a regional line of enterprise software which might actually be a good fit, right? So all that stuff's up there, and readily available to everybody, so the key issue is deciding which document is most like the one you are about to write.

A manager from another firm emphasised that creativity requires reusing previously written materials in a 'good way'. 'To a certain extent, this can be a challenge to your writing skills', she said. 'You can't simply copy-and-paste. You have to think about the effective pitching angle'.

One major theme that ran through our discussions and observations, was that creative texts were not much use without a good creative idea at the foundation of a given campaign, and good ideas that addressed particular goals for particular clients at particular points in time were much more difficult to reuse. One manager said, 'Sometimes people would go and cut-and-paste something and they would get it wrong, because they're not only cutting-and-pasting the structure and the content but also the creative idea, which you know may not be appropriate for every context'. Another described creative processes as existing in a kind of hierarchy:

> It's sort of the hierarchy of creativity, right? So, at the top is the campaign theme and it has to be something clever, twisted or it makes the target audience think about something differently and behave differently, and then below that are the creative tactics, and then below that there's the creative content which often involves doing things that are actually fairly ordinary, so what determines whether something is creative at the bottom of the food chain depends on that central strategic or interesting thing that's decided at the top.

Interaction orders

All of the stages in the creative hierarchy described above involve some kind of collaboration. Formulating creative strategies often takes place in group brainstorming sessions; tactics are designed and implemented by teams, and texts are often composed collaboratively. This is, of course, not surprising. The literature on professional communication has consistently shown that collaboration is the norm rather than the exception in most industries, including the creative industries (Bilton, 2007; Cross, 2001; Ede and Lunsford, 1990). As one senior vice president of a regionally based public relations firm told us:

> Business is all about working in teams and PR is really all about project management . . . but I can really say that from my experience of working for small and medium (PR) enterprises over the past decade or more there is such a variety of teamwork and in truth some is more effective than others to do the job.

So what kinds of *interaction orders* are the most effective for facilitating creativity? Ede and Lunsford (1990), in their classic work on collaborative writing, identify two modes of collaboration: 'hierarchical collaboration' in which the relationship between collaborators is unequal, and 'dialogic collaboration' in which the relationship is more egalitarian. This distinction, however, is in some ways problematic in that it implies that 'dialogue' is a function of egalitarian rather than hierarchical face relationships;

our findings, however, indicate that the situation is actually much more complicated, with hierarchical relationships sometimes equally useful if not more useful in creating an atmosphere of open and equal participation and free exchange of ideas.

In any case, as in all business contexts, hierarchy is a given in public relations firms. All participants operate within a web of hierarchical relationships. One general manager of a large multinational firm, in fact, described the creative process as 'a process of working up and down through hierarchies within the office'. Moreover, these hierarchies were often complex and multidimensional, with power accruing to different participants at different times for different reasons. He said:

> You've got these hierarchies all over the place, within teams, between teams, you've got different teams and different offices working together, and then you've got the client–agency dimension going on as well, and so it's layer upon layer of this interconnectivity.

Against the backdrop of these complex networks of hierarchy, participants often found it necessary to negotiate different kinds of face relationships at different stages in the creative process. Brainstorming sessions, for example, were seen to work best when everyone felt that they had an equal right and an equal responsibility to contribute, whereas in the stage of drafting documents more rigid and hierarchical role differentiation was favoured.

Ironically, however, it was often the underlying hierarchy which often seemed to foster more creative participation. In brainstorming sessions, for example, team leaders and other more powerful participants saw it as part of their job to create the space for less powerful people to contribute, and also to ensure that more aggressive or talkative members did not dominate the discussion. The general manger of a medium-sized regional firm described the process like this:

> You have to give them permission, to have an idea, right? Sometimes the most quiet person in the room will be the one who has the most creative and interesting thing to say, but they have to be given the permission to have a good idea, and you have to be very sensitive to people's place in the organisation and hierarchy and understand how that's going to affect what and how they articulate.

Other senior staff we interviewed discussed ways in which they encouraged participation through practices like giving credit to junior employees for successful group ideas and making use of various group activities and written exercises to encourage less powerful members of the team to be more assertive. In this regard, previously authored documents (*discourses in place*) often played an important role. One owner of a small local firm, for example, was in the habit of passing out brainstorming lists or mind maps from previous campaigns to participants before meetings in order to stimulate their thinking.

This process of enacting egalitarian interactions within hierarchical relationships, however, was not without difficulties, the most obvious being that the symmetrical face relationships and assertive behaviour expected in brainstorming meetings often contradicted the kinds of face relationships enforced in most other situations within the firm. Some of the junior staff we talked to felt very keenly the performative or ritualistic nature of these 'egalitarian' brainstorming sessions. 'Yes, we try to reach a consensus and vote', said one consultant, 'but in the end the final decision rests with the boss. Since he's the boss, we all agree with him'.

Hierarchical relationships are much more explicitly enacted in other aspects of the creative process, most obviously in the drafting of documents, which is almost always done in hierarchical dyads in which the division of labour is usually very clearly laid out. In some situations we observed, junior staff conducted research and gathered information while more experienced staff drafted the documents, and, in others, less experienced staff drafted documents, and more experienced staff revised and approved them. Often, a combination of these two patterns was seen, with the frequency of each pattern varying over time. Brand new staff were rarely given the responsibility for writing documents and spent most of their time on research. After their more experienced colleagues or supervisors had modelled the drafting of documents a number of times, however, these junior staff gradually moved on to do the drafting themselves (often with even less experienced staff or interns assigned to gather information for them). A consultant in a medium-sized firm described the process this way:

> Me as a junior in the team, I'm actually the person writing, researching the article, and putting it all together. And my boss is actually the team leader . . . he is the person who gives comments on the drafts . . . and basically tells us the revisions we need to make in order for it to go out.

Interestingly, these more explicitly hierarchical arrangements often seemed to better facilitate individual inventiveness, initiative and genuine dialogue between participants than the less explicit hierarchies of brainstorming meetings. Individuals working alone could focus on coming up with new ideas without having the extra burden of having to manage the rather fuzzy and contradictory face relationships associated with group brainstorming, and, in the best of circumstances, more powerful members of such dyads could support and scaffold creativity in more direct ways. Another consultant noted, 'We have a lot of these press releases, so actually our superior gives us a lot of flexibility and gives us a lot of room to do our own thing'.

In fact, we often observed these hierarchal dyads developing into mentoring relationships in which the focus of the superior's feedback was as much on developing the junior staff member's writing ability as on improving the draft. An entry in our research assistant's reflective diary describes one such relationship like this:

> In this meeting, (supervisor's name) and I sat down together to talk about the changes he made on the draft, and he explained the reasons why he had made these changes. He gave many useful suggestions and comments . . . In terms of language use, he focused a lot on the words, pointing out which words I had been overusing and suggesting some more specific words . . . As for content, he added several constructive suggestions in the '*Notes to Editors*' of the press release. He also gave positive feedback and encouraging comments, 'I was so happy and impressed by this. If I had to grade this, I would give it an A+!'

Of course, such relationships do not always grow out of the hierarchical face systems associated with the drafting process. The managing director of a medium-sized firm, for example, related her past experiences working in much less successful hierarchal dyads:

> I've worked for PR bosses who ask you to write a press release and seem to take great delight in telling you that it sucks without indicating how it can be improved and then you go through endless rounds of time wasting trying to get it right until the deadline looms and the boss does it themselves, scrapping all of your work. But even worse, the release then goes to the client who sends it back full of edits and nasty comments about it being useless and then your boss blames you again and you're back at the start of a time wasting cycle.

Historical bodies

From the perspective of *discourses in place*, creativity is seen as primarily a question of resources. From the perspective of *interaction orders*, it is seen primarily as a question of the successful management of hierarchical relationships. From the perspective of *historical bodies*, creativity is chiefly seen as a matter of professional development. Paradoxically one of the main complaints of the managers and other senior staff we interviewed was a perceived lack of sufficient creativity and initiative among their employees, and much of what they described as creative processes within their firms were focused not just on producing creative texts or creative campaigns but also on ways of producing creative people. Most of the senior staff we spoke to saw their employees as very much 'works in progress'. One managing director, in fact, described her company as a kind of 'finishing school' for creative professionals.

Clearly, the interaction orders described above, characterised by hierarchical dyads, played an important role in helping to socialise new staff into the creative practices of the firm. One managing director of a medium-sized regionally based firm described her own experiences of this kind of creative apprenticeship as follows:

> I won't deny that it was hard at first and I was shy about contributing to the discussions of internal and client meetings and even writing news releases took me ages at first, but I had a great managing director who kept reminding me, 'Look, the more you do it the better you'll get – you'll be writing releases in your sleep like you had a PhD in it!' and she was right, and this is a saying that I use to motivate my juniors and new hires today.

Much of this learning, of course, takes place in the context of negotiating creative ideas and creative texts. 'I learn a lot when people edit my work,' said one consultant. Another way this training takes place is through less experienced staff 'shadowing' and observing their seniors. 'I think that you really learn a lot by sitting next to someone and listening to them pitching and giggling with the media', one account executive said, 'you know, showing off, doing your PR flirt, whatever that might be'.

Discourses in place, in the form of templates and boilerplate, were also seen to play an important role in the socialisation of newcomers into a firm's particular creative practices, and thus in helping to promote and maintain the creative culture of the organisation as a whole. One consultant we talked to put it this way:

> Yes, I think this is very important. It's the corporate culture of our company. As I said before, the turnover rate in the PR industry is quite high. There are lots of new staff coming to our company every year. These templates are important for maintaining the style, or the corporate culture, the image of our company.

Related to the circulation of previous written documents was the frequent circulation of documents and drafts among all team members *as they were being prepared*, whether or not the recipients were directly involved in drafting or editing the document. A senior account executive described this practice of distribution like this:

> I guess it's another form of cooperation, how we internally communicate. For any team that's working with a client there's always a cc on every single communication. So we'll cc all of them and they'll cc all of us so you know any time everything that's going on . . . Within the life of a press release or something, a written document, there would be a cc when they send to the copywriter. I would see the draft come back from the copywriter and there would be a cc from the account manager to myself. Um . . . so every level of work there is this kind of shaded involvement and I think it's a very important way of training people.

As practitioners move though these processes of formal training and less formal socialisation, supported by *discourses in place* and the *interaction orders* that provide opportunities for what Lave and Wenger (1991) call 'legitimate peripheral participation', they gradually move from more novice participation to positions of expertise in which they take responsibility for training and mentoring younger staff and for contributing texts to the firm's archive of discourse.

Conclusion

Based on the analysis above, creativity in the public relations firms we studied can be defined as a coordinated process of problem-solving achieved through the appropriation, adaptation and mobilisation of discourse across time and space. In the course of public relations campaigns, discourse moves across multiple modes (Iedema, 2001) from oral conversations in brainstorming meetings, to written texts such as proposals, press releases, pamphlets and web pages, and sometimes to more embodied modes like speeches, events and media appearances. Each of the different stages in a PR campaign involves different resources for creativity, different sets of social relationships among those involved, and different opportunities for individuals to develop their own creative potential. In other words, each stage in the process constitutes its own *nexus of practice*, with *discourses in place*, *interaction orders*, and *historical bodies* interacting with one another in particular ways. In brainstorming sessions, for example, discourses in place are often used to scaffold more egalitarian interaction orders in which individual staff gain experience in debating and negotiating ideas and reaching a consensus. In the drafting of documents, discourses in place, in the form of templates and boilerplate, circulate through the process, supporting more hierarchal interaction orders in which expert writers assist novice writers in both showing off their individual abilities and mastering the creative practices of the firm. One of the main challenges firms face when it comes to facilitating creativity is being flexible enough to accommodate these different configurations of discourses, interaction orders and individuals associated with different kinds of creative activities.

As with the kind of creativity discussed in the previous chapter by Bhatia, creativity in public relations firms is not so much textual as *intertextual* and *interdiscursive*, located less in discourse itself and more in the links between different discourses (see also Bhatia, 2010). In examining these intertextual and interdiscursive links, however, we should not just focus on the links among different texts, but also explore the links that discourse processes help to build among people, between individuals and the teams they are part of, and between people's past experiences and their future potential.

It is perhaps easy to be dismissive of creative processes in commercial concerns such as public relations firms, especially when viewing them from the point of view of literary or artistic creativity, the standards of which the sometimes formulaic texts of public relations, often cobbled together with bits and pieces from previously authored documents, can hardly live up to. Even public relations practitioners themselves, in fact, tend to be deeply cynical about the whole notion of creativity in their profession (Green, 2010). For successful public relations firms, however, creativity is not primarily a matter of individual texts, or even of individual campaigns, but rather a broader matter of professional development and institutional growth. The creative recycling and resemiotisation of previously authored texts in the firms we studied was not just a way of utilising resources more efficiently or economically. The cycles of discourse set in motion as texts circulated from campaign to campaign, through the various stages within campaigns, through the interaction orders of various teams, and through the historical bodies of various practitioners, were central to both sustaining the creative cultures of the organisations and to facilitating the creative development of individuals.

Note

[1] Research reported in this paper was supported by an RGC GRF grant from The Government of HKSAR [Project No. 9041281 (City U 143807) *Collaborative Writing in the Creative Communication Industries: Professional and Pedagogical Perspectives.*]

References

Aronson, M., Spetner, D. and Ames, C. (2007) *The public relations writer's handbook*. San Francisco: Jossey-Bass.

Bakhtin, M. M. (1991) *The dialogic imagination: Four essays* (ed. M. Holquist). Austin: University of Texas Press.

Bhatia, V. K. (1993) *Analysing genre: language use in professional settings*. London: Longman.

Bhatia, V. K. (2010) 'Interdiscursivity in professional communication'. *Discourse and Communication* **21** (1) 32–50.

Bilton, C. (2007) *Management and creativity: From creative industries to creative management*. London: Wiley-Blackwell.

Cross, G. A. (2001) *Forming the collective mind: A contextual exploration of large-scale collaborative writing in industry*. Cresskill, NJ: Hampton Press.

Ede, L. and Lunsford, A. (1990) *Singular texts/plural authors: Perspectives on collaborative writing*. Carbondale: Southern Illinois University.

Green, A. (2010) *Creativity in public relations* (4th edn). London: Kogan Page.

Iedema, R. (2001) 'Resemiotization'. *Semiotica* **137** (1–4) 23–39.

Lave, J. and Wenger, E. (1991) *Situated Learning: Legitimate peripheral participation*. Cambridge: Cambridge University Press.

Norris, S. and Jones, R. H. (2005) *Discourse in action: Introducing mediated discourse analysis*. London: Routledge.

Reddi, V. V. N. (2009) *Effective public relations and media strategy*.

New Delhi: PHI Learning Pvt. Ltd.

Scollon, R. and Scollon, S. W. (2004) *Nexus analysis: Discourse and the emerging internet*. New York: Routledge.

PART THREE

Multimodal creativity

The very inclusion of this section in a book on discourse and creativity reflects a profound shift in the study of discourse in the past two decades from being 'logocentric', that is, concerning itself almost exclusively with written and spoken language, to seeing all meaning-making as essentially 'multimodal', involving not just words but also other semiotic modes such as image, gesture, music, object handling and architectural layout. Along with this acknowledgment has come the exploration of how analytical tools that have been developed for the study of discourse in its linguistic form could be applied to the study of non-linguistic aspects of texts. This exploration has led to the development of a number of analytical frameworks for what has come to be known as 'multimodal discourse analysis', three of the most prominent of which are represented in this section. The first chapter by Forceville on visual metaphor in advertising demonstrates an approach to multimodal discourse which draws heavily on the cognitive sciences in general and in particular on work within the field of cognitive linguistics. The chapter by van Leeuwen demonstrates through an analysis of jazz music what has come to be one of the most influential approaches to multimodal discourse, an approach based on the principles of social semiotics. Finally, the chapter by Norris illustrates a more interactional approach to multimodal discourse influenced by interactional sociolinguistics and mediated discourse analysis.

Not surprisingly, these different approaches to multimodality result in different ways of defining and analysing creativity. For Forceville, creativity is very much a matter of cognitive processes in which previously unconnected concepts or domains of experience are brought together in particularly useful, concise and elegant ways. Using an enhanced version of Fauconnier and Turner's 'Blending Theory', which provides a framework for understanding how 'human beings constantly combine relatively simple concepts in order to arrive at more complex ones', he shows how advertisers creatively use images to create similarities between different domains or 'mental spaces' in ways in which the positive attributes of one domain are transferred to the other domain. His discussion, however, goes beyond describing how images create similarities between domains to considering why visual metaphors seem to be more effective in advertising than verbal ones due to their ability to enable meanings which cannot be rendered through words alone and the fact that they demand from the 'reader' more mental work in construing the metaphor and forming inferences about it.

While Forceville's view of multimodal creativity focuses on cognitive processes, van Leeuwen's focuses on the *social* processes around the design, production and distribution of creative artifacts. Creativity for him has to do with the ways in which people situate themselves within particular traditions of meaning-making as they make use of the semiotic resources that are available to them in their particular sociocultural contexts and, sometimes, invent new ones. From this perspective, the role of the discourse analyst is not just to explain what makes a particular text or combination of semiotic resources 'creative', but to understand how creativity itself 'drive(s) semiotic change in particular cultural and historical contexts'. Through his examination of the evolution of jazz music, van Leeuwen shows how various design schemata, aspects of production, and modes of distribution have combined in different ways over the years to open up possibilities for the invention of new resources for meaning-making. One particularly interesting aspect of van Leeuwen's historical perspective is his take on post-modernity as a kind of paradigm shift in our understanding of creativity from the creation of new design schemata to the remixing of old ones using new tools for production and distribution. In this new paradigm, he argues, the creator has become a kind of technician, with the resources for design and production increasingly constrained by the options and algorithms that are built into the software we use.

The final chapter of this section by Sigrid Norris takes a very different approach both to the analysis of multimodal discourse and to the analysis of creativity. Norris is less interested in the formal aspects of 'creative' texts or with the ways the different semiotic resources available in a particular domain of creativity change over time, than she is with the social construction of the 'creative person', and how this construction is achieved moment by moment through the appropriation of various multimodal resources into everyday interactions. How is it, she asks, that some social actions and social actors are construed as 'creative' and others are not, apart from aesthetic evaluations of the artifacts they produce? Through an ethnographic analysis of the evolution of the identity of a particular community artist, she shows how the construction of the creative person is a long-term social process of negotiation between individuals and the social networks to which they belong, mediated not just through artistic products but through a whole range of meditational means including living environments, tools such as paint brushes and canvasses, and personal narratives about the self.

CHAPTER 7

Creativity in pictorial and multimodal advertising metaphors

Charles Forceville

Introduction

'The logical pattern of the creative process', declares Arthur Koestler in his classic study, 'consists in the discovery of hidden similarities' (1969: 27). Discussing numerous examples he demonstrates how this creative act of unveiling similarities functions in the realms of humour, scientific discovery, and art. Koestler coins the term 'bisociation' for the unexpected fusion of two elements: 'The bisociative act connects previously unconnected matrices of experience' (1969: 45). The fusion of the two objects, phenomena, or perspectives, then, results in the production of something new. But this in itself is not yet sufficient to merit the qualification 'creative'. Human beings' behaviour is fundamentally goal-driven. Consequently, creativity must always be assessed and evaluated within a more or less specific situation in which a 'maker' (of a sentence, a building, a poem, a song, an argument, a painting, an advertisement, a software program . . .) searches for, or hits upon, a solution to a more or less clearly delineated problem she encounters – or has formulated herself. The joke must be funny, the scientific explanation correct or insightful, the artistic object felicitous or striking. Moreover, the result must reflect a principle of 'economy' (Koestler, 1969: 263): it must be as concise and elegant as possible.

Due to the highly contextualised nature of creativity it is virtually impossible to come up with non-trivial generalisations about it. But perhaps the difficulty is even more fundamental: generalising about creativity entails finding pertinent patterns in pattern-breaking. This is a tall order. The best I can do here is examine and analyse a number of clearly goal-driven situations within a specific genre (see Forceville, 1995) that result from bisociation in roughly the same manner, and evaluate and discuss the solutions found.

Finding their inspiration in Koestler's work, Fauconnier and Turner have developed a theory, Blending Theory, which among other things can model 'emergent structure' (2002: 383), and hence potentially 'creativity.' Although Blending Theory suffers from certain shortcomings (Forceville, 2004; Câmara Pereira, 2007: 53), I agree with Câmara Pereira that it can help formalise creative processes. In the following sections I will first briefly explain Blending Theory, and then specify, in terms defined by Veale *et al.* (in prep.), how the Blending Theory model needs to be enhanced in order to be truly capable of handling creativity. Finally, some supposedly creative metaphors in advertisements and commercials will be discussed in terms of this enhanced Blending Theory model.

It is to be acknowledged at the outset that the attribution of creativity is *post hoc*, a kind of 'reverse engineering', which will not yield any spectacular predictions about future creativity. Nonetheless, I believe it is useful to analyse specimens of pictorial and multimodal metaphor in a specific genre from the perspective of their creativity. Such case studies alert us to how the choice for specific modalities (here: spoken and written language, visuals, sound, and music) affects a metaphorical bisociation's successfulness. Indeed, the case studies examined here will show that often the creativity of the solutions chosen resides not so much in the *content* of pertinent information (which in advertising is often predictable or even banal), but in conveying this information in unexpected *form* – which in the present cases means in non-verbal or multimodal ways. This issue should be of interest both to analysts and producers of advertising metaphors.

Blending Theory, Interaction Theory and Creativity

Fauconnier and Turner's Blending Theory (Turner and Fauconnier, 1995; Fauconnier and Turner, 1998, 2000, 2002; see also Grady *et al.*, 1999), is rooted in Conceptual Metaphor Theory (see Lakoff and Johnson, 1980, 1999; Gibbs, 2008), which in turn is a prominent branch of a brand of linguistics usually labelled 'cognitive linguistics'. Over the past two decades, the cognitive linguistic model has begun to be used for the analysis of non-verbal modes of discourse, such as gestures, visuals, sound and music

(Forceville and Urios-Aparisi, 2009). Cognitive linguistics thus contributes to the theorisation of the quickly developing discipline of multimodal discourse along lines that differ from those proposed by another prominent model, the Hallidayan Systemic Functional Linguistics paradigm, as for instance exemplified in Kress and Van Leeuwen (2001) and Jewitt (2009). Assessing the strengths and weaknesses of the two linguistics models as instruments for analysing multimodal discourse will require sustained research and will be outside the scope of this chapter (for some thoughts, see Forceville, 2011).

Blending Theory provides a model claiming to describe 'the way we think'. While assessing the import of this ambitious claim will have to await both more theoretical discussion and empirical research, Blending Theory at the very least offers a plausible template for modelling how human beings constantly combine relatively simple concepts in order to arrive at more complex ones. One of the phenomena Blending Theory can model is metaphor. In as much as it is capable of modelling creative metaphor, it can be seen as fleshing out Max Black's (1979) interaction theory (also known as the 'creativity theory') of metaphor, which in many ways anticipates Conceptual Metaphor Theory.

Let me first briefly recapitulate Black's theory. Black distinguishes between a metaphor's primary subject (for which I will here adopt the label 'target') and a secondary subject (here: 'source'). In the metaphor 'a good manager is a conductor', 'manager' is the target, and 'conductor' is the source. Each metaphor functions by inviting the processing, or spelling out, of one or more features that are to be projected, or 'mapped', from source to target. That is, the addressee of a metaphor selects one or more features metonymically connected with, or associatively evoked by, the source. Since the source (just as the target) is not a monolithic concept but part of a semantic domain, the addressee will look for features in the source domain that, in the given context, can be sensibly mapped on to the target domain. In our example metaphor, such mappings may include for instance that it is the conductor who leads the orchestra without him or herself playing an instrument and that it is s/he who indicates the moment when an instrument or group of instruments begins to play, as well as the tempo in which this is to be done. In the absence of context that may further steer and constrain interpretation, we may decide on other mappings: the conductor's artistry, for instance, and his/her 'star' status. Of course the mapping involves a degree of 'translation': the musicians in the source become employees in the target; 'playing an instrument' becomes 'doing your job', 'deciding on the moment the violins need to start playing' corresponds to, say, 'deciding on the moment the marketing department is asked to launch a product'.

A few insights transpire from this example: in the first place, if the textual or discursive context does not specify what elements from the source domain are to be mapped, and how they end up after their translation in the target domain, addressees have a degree of freedom to interpret the metaphor. This means that – again: in the absence of constraining contextual information – there may be divergent interpretations of the metaphor that are all correct, but may nonetheless lead to disagreements between interpreters who may not realise that their dispute stems simply from having mapped different features from source to target. In the second place, not only factual properties can be mapped, but also emotions, connotations, and valuations, such as, in our example, 'artistry'. Thirdly, it is also clear that not every feature can be mapped; in that case target and source would have to be identical, and the result would be a tautology rather than a metaphor. For instance, the fact that the conductor uses a baton does not have an equivalent in the target domain – unless one were to label the chairman's gavel as such.

Black's metaphor theory can be conveniently modelled using the Blending Theory template given in Figure 7.1. Here is an explanation of a simplified version of that diagram, the understanding of which will be aided with reference to Fauconnier and Turner's own example of the blend 'land-yacht' for a very big, imposing car. Two 'mental spaces' ('small conceptual packets constructed as we think and talk, for purposes of local understanding and action', Fauconnier and Turner, 2002: 40) each contribute salient factual, emotive, and/or associated information to the blend. In our example the mental spaces of 'yacht' and 'car' (the two 'input spaces') help build the metaphorical blend 'land-yacht'. They can do so because the input spaces on the one hand have certain *shared* salient features, or structured aggregates of features (what Gentner calls 'relational alignment', e.g. Holyoak *et al.*, 2001: 8); and on the other hand they have certain *non-shared* salient features or structured aggregates of features. Both the shared and the non-shared features are necessary for the blend. Only on the scaffold of a partial commonality between the contributing input spaces (Koestler's 'matrices') can a creative pertinent fusion arise. The shared salient features here include 'constituting a vehicle for the transportation of people', 'being equipped with a motor', 'costing a robust amount of money', etc. Features that the yacht space does *not* share with the car space are 'being typically propelled by wind', 'having a mast' and, importantly, 'being luxuriously big and stately' and 'used primarily for leisure purposes'. Conversely, the car space does *not* share with the yacht space features like 'moving forward on four tyres' and, importantly, 'using *roads* as the prototypical trajectory'. The features the two input spaces have in common are represented in the so-called 'generic space' – and it is these (aligned) features that enable the two input spaces to be experienced as similar. In the land-yacht example this could be summed up as 'being a vehicle for the transportation of people'. But since the input spaces are non-identical in at least some respects, they can also impart features to the blended space that are not shared by the other input space. Indeed, they *must* do so: each contributing space will at the very least impart one feature to the blend that it does not share with the other input space(s) – or else the rationale to deploy that input space for contributing to the blend has disappeared. The resulting whole, if it is to count as 'creative', is more than the sum of the parts or, as Fauconnier and Turner succinctly phrase it, 'the blend develops emergent structure that is not in the inputs' (2002: 42). In the land-yacht example, the car space imparts the feature 'travels by road', while the yacht space imparts 'being luxuriously big and stately' and 'used primarily for leisure purposes'. The emergent structure in the blend saliently includes, I propose, unsuitably and intimidatingly large size (as said, for instance, of a Hummer) – but different users may map different valuations and connotations from the 'yacht' space.

In Figure 7.1, these processes are formally represented as follows. The big circles are mental spaces. The left and the right space are the input spaces, while the space at the top is the generic space, representing what the input spaces have in common. At the bottom is the blended space, which

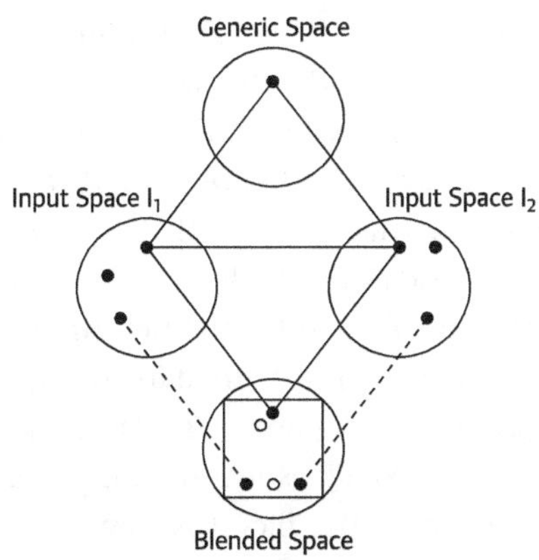

FIGURE 7.1 Fauconnier and Turner's Blending Theory
(adapted from Fauconnier and Turner, 2002: 46)

symbolises the *ad hoc* combining of pertinent properties from the two input spaces into a new whole. The black dots represent properties. The uninterrupted lines between dots represent a property shared across input spaces. Since this property is shared, it naturally also occurs in the generic space, which is why there are uninterrupted lines connecting the dots in the input spaces with a dot in the generic space. The same property is also retained in the blend, which is again represented by uninterrupted lines running from the input spaces to the dot in the blend. The interrupted lines, those between dots in the input spaces and dots in the blend, indicate properties uniquely imparted to the blend by each of the input spaces. The square in the blended space contains the pertinent properties in the blend. Since not all properties from the input spaces are imparted to the blend, there are some isolated, unconnected dots in the two input spaces: these are properties not relevant for the blend. The open, as opposed to the black, dots in the blend signify properties that are born thanks to the combining of the input spaces: these, then, symbolise the new, 'emergent' properties.

Rephrased in terms of Black's metaphor theory, input space 1 is the original, 'unadapted' target domain ('car', 'manager'); input space 2 is the source domain ('yacht', 'conductor'); and the blend is the 'transformed' target domain ('land-yacht', 'manager-as-conductor'). The generic space labels the pre-existing similarity between target and source, which is retained in the blended space (e.g. car and yacht both have motors; they are both vehicles; and manager and conductor are both human; they are both in charge of a group of people who are to carry out a task together, etc.).

Inasmuch as the transformed target/blend provides properties (and the structure binding those properties) that were present in neither the unadapted target domain nor the source domain, it is correct to talk about 'emergent structure' (see Gineste *et al.*, 2000). In principle, that is, the Blending Theory is capable of modelling what happens in creativity. As argued in Veale *et al.* (in prep.), however, this requires some further specification. Briefly, in a creative act (whether of the radically innovative variety of the 'never done before in history' called 'H-creativity' (where the 'H' stands for 'historical') or of the more local variety of 'I hit upon this, to me, completely new discovery or solution for the first time', called 'P-creativity' (P = 'personal') by Boden (2004: 2), the unadapted target (input space 1) is given, and so is the more or less vaguely formulated outcome of the 'transformed target/blended space'. What is given in input space 1 is: 'car' and 'manager'; what is to end up in the blend is, say: 'a felicitous description of an inappropriately big type of car' and 'a felicitous description of a manager emphasising his artistic qualities', respectively. But what makes the blend creative is precisely the contents of input space 2 or source domain: the yacht, the conductor – and this is *not* given in advance. Finding an appropriate source domain whose properties can merge with those of the target in order to create a felicitous metaphorical blend – that is the big challenge for any would-be creator. In Veale *et al.*, this problem was formulated as follows: 'The mechanism by which the contents of a second input space [in metaphor: the "source domain", ChF] are inferred to complement the contents of the first [in metaphor: the "unadapted target domain", ChF] is the mechanism that will make blending theory a true theory of creativity' (Veale *et al.*, in prep.). Note that the fact that the transformed target/blend contains a description (necessarily vague because it still lacks the crucial information from the source) already indicates that the producer of the blend has a sense of purpose or goal. This is a reminder that a semantic analysis of a blend must always be complemented by pragmatic considerations.

Creativity in advertising metaphors: case studies

Advertisers' self-imposed task for creative pictorial/multimodal metaphor use, I propose, can be formulated as follows: 'Develop a creative representation of a semantic domain (source domain) that succinctly and surprisingly, perhaps humorously or beautifully, cues pertinent features and/or structures that can be mapped onto the product promoted (target domain) so as to present the product as exemplifying the feature(s) that is/are to be communicated to the audience'. This task, which in the last resort banally boils down to 'make 'em buy!', can be, somewhat artificially, divided into the following subtasks:

1. Decide on the attributes and/or connotations of the product/service to be emphasised.
2. Find a source domain in which those attributes are saliently present.
3. Create a motivation or scenario for the co-presence of target (usually: the product, Forceville, 1996: 126) and source, and emphasise dimensions of similarity between them.
4. Choose in what modes/modalities the metaphorically pertinent information (including the to-be-mapped properties/attributes) is to be cued.

In the following I will discuss a number of Dutch examples with these questions in mind.

Artis Zoo

The goal of the billboard in Figure 7.2 (Forceville, 1996, 158 ff. Original in colour) is to attract visitors to the Amsterdam zoo 'Artis'. The source domain chosen to emphasise Artis' visitability is Leonardo da Vinci's world-famous painting 'Mona Lisa' (Figure 7.3 overleaf). This portrait of a woman displaying an enigmatic smile, of which many people are supposed to have a mental picture, is used to suggest a similarity with the photographic portrait of an orang-utan that also appears to be faintly smiling. This 'portrait' framing, combined with the smiling, creates the pertinent similarity between target and source. This newly created similarity between two phenomena (orang-utan and Mona Lisa), giving rise to the metaphor ORANG UTAN IS MONA LISA, spawns further similarities, namely between the two institutes that they metonymically represent (Artis Zoo and the Louvre). Thus, the Louvre's property 'famous tourist attraction that people want/should/need to visit' is mapped on to the Amsterdam zoo. As to the choice of modes (see Forceville, 2006, 2008; Bateman, 2008; Kress, 2009, 2010 for more discussion on the concept of 'mode'), these comprise visuals and language. The decision to render the source domain 'Mona Lisa' verbally rather than juxtaposing it visually to the orang-utan works out well; presumably, once viewers realise what metaphor they are to construe, they will adapt their mental picture of the Mona Lisa to optimise the similarity with the monkey, whereas if one were to see a picture of the Mona Lisa next to the monkey (which for copyright reasons may not have been an option for the advertisers in the first place) would have resulted in a less convincing similarity. A final remark: this particular orang-utan was one of many possible target domains that could have been chosen (although another billboard in the same Artis series showed a lovely baby chimpanzee with the text 'King Kong?', which might suggest that the designers of the campaign opted for a 'monkey' theme). The *formal* similarity between the orang-utan and Mona Lisa that resides in both being portraits of faintly smiling primates – even if the similarity did not exist before this billboard was conceived – is an important factor in the metaphor's success.

Nespresso (print ads, 2009, originals in colour)

From the tag line 'Citiz: high design by *Nespresso*' (pp. 123 and 124) we can deduce that the design of the coffee machine, rather than for instance its price or the taste of the coffee it produces is the attribute to be promoted in this advertising campaign. The visual context reinforces this impression,

FIGURE 7.2 Billboard for Artis Zoo, Amsterdam, 1990s
(photo © Charles Forceville)

FIGURE 7.3 Leonardo's Mona Lisa, Louvre, Paris
(photo © Classic Image/Alamy)

very strongly suggesting that the three machines of the Citiz line are inspired by modern urban architecture (Figures 7.4 and 7.5). The machines are seen against the background of metropolitan skyscrapers, reminiscent of the Parisian La Défense or the Twin Towers. The similarity between the robust, metallic machine and the accompanying coffee canister on the one hand, and the skyscraper suggest New York more specifically. Clearly, then, it is the design attribute itself that is to be mapped. The realistic motivation of the co-presence of target and source is presumably the scenario that the machine stands on the edge of a skyscraper balcony, or on a ledge before a big glass window. Note that the notion of 'glamour' may be co-mapped together with 'state-of-the-art design' from cityscape to coffee machine; this notion also comes to mind when one considers the endorsement by actor George Clooney in the campaign promoting Nespresso coffee during this same period.

In my terminology (Forceville, 1996, 2006, 2008), this is a pictorial metaphor (more specifically of the simile type). That is, although the advertisements' texts and the machine's name ('Citiz') help cue the source domain, the visuals alone suffice to construe the metaphor COFFEE MACHINE IS SKYSCRAPER. It makes sense that the visual modality is dominant: attesting to excellent design after all is something to be shown rather than verbalised.

FIGURE 7.4 Print ad for Nespresso coffee machine
(© Nestlé)

FIGURE 7.5 Print ad for Nespresso coffee machine
(© Nestlé)

Senseo coffee machine (commercial)

Presumably the goal of this 2004 commercial, shown on Dutch television, was to reinforce the popularity of the brand, a Philips coffee machine that had already been enormously successful in the first three years of its existence. The source domain to be found for its metaphorical promotion thus had to connote, at the very least, 'popularity', more specifically the kind of popularity that has achieved cult-status. The source domain chosen to cue this 'legendary status' is that of *Easy Rider*-type motorbikes. This is done as follows: the motorbike domain is triggered by a number of visual close-ups of elements that are metonymically related to this domain: metallic parts, a finger pushing an on/off button, a drop of oil – or so it seems – accompanied by a vroom-and-kick-starting sound. At this stage, first-time viewers would be unaware that what is triggered is in fact a source domain that will metaphorically predicate something about a hitherto undisclosed product (i.e. coffee machine), rather than a promotion of motorbikes. The motorbike domain is further enriched by the song audible on the soundtrack: Steppenwolf's 'Born to be Wild', the opening song of the cult film *Easy Rider*.

In this 1969 movie, the characteristic motorbikes that Dennis Hopper and Peter Fonda ride on, and the whole atmosphere of this very loosely structured film, is the 'legend' metaphorically exploited in the commercial. The following texts are superimposed over the images in the course of the 30-second commercial: 'Designed with a vision . . . developed with passion . . . turns each moment into a sensation . . . that strokes all senses. . . .' – all descriptions that nicely fit the *Easy Rider* motorbikes. Not until the very end of the commercial (embedded in Forceville, 2007) do viewers realise they have been tricked: the metal parts and the on/off button were actually part of the coffee machine we now get to see ('three years old, and already a legend . . . at least for the kitchen'), while the 'drop of oil' was in fact, they retrospectively realise, a drop of coffee. The formal similarity is thus cued by visual means, while the desired, mappable 'legendary' attributes – something like 'the freedom, sixties anti-conformism, youth-culture, etc. associated with *Easy Rider*' – are cued by the Steppenwolf song.

The similarity between target and source is thus completely commensurate with Max Black's interaction theory: it is *created* rather than pre-existent. Outside the context of this commercial, nobody would have thought of the similarity between the Senseo coffee machine and hip motorbikes, let alone the motorbikes of *Easy Rider*. The choice of modalities to capture the similarity between target and source is effective. The pay-off 'already a legend' captures this similarity verbally, but up until that moment in the commercial, the 'legendary' aspect was for the audience to construe. Inserting footage from the film would not, I submit, have worked: simply juxtaposing the machine and the *Easy Rider* bikes to convey the former's 'legendary' status would have been a ludicrous overstatement. By contrast, the clever close-ups combined with the song cue the metaphor SENSEO COFFEE MACHINE IS *Easy Rider* MOTORBIKE implicitly rather than explicitly. The audience (at least the older part that is aware of the song and the association with the film – undoubtedly the target audience of the commercial) has to *infer* the connection. It is to be noted that viewers' interest and their sense of surprise are enhanced by being presented with the source *before* the target: a choice that appears to be a standard procedure for metaphors in commercials (see Forceville, 2007). Finally, the metaphor is presented in a tongue-in-cheek spirit by the addition 'at least for the kitchen' in the pay-off.

Zendium toothpaste (commercial)

After a colourful shot of bubbling water we see beautiful white coral, then we notice the same white coral affected by black spots – now in a black and white shot. Zendium toothpaste containers swim with fish-like movements in an underwater seascape – in colour again – toward the coral. The voice-over text runs, in translation, 'Saliva, it is the natural protection of your mouth and teeth. If the condition of your saliva deteriorates, mouth and teeth problems may arise. Zendium is the only toothpaste that strengthens your saliva so as not only to protect your teeth, but your entire mouth'. The last phrase sums up the product's central claim to fame and thus the attribute to be emphasised. In order to achieve this, a metaphor is construed that could be labelled SEASCAPE IS MOUTH. Thanks to the voice-over, we associate the white coral with teeth, and the black spots on it with caries. We do not need the voice-over to understand that the Zendium containers are fish; that can be inferred from the containers' fishlike movements. What matters is that just as the natural habitat of fish is water, so Zendium's natural habitat is saliva, and this is the salient mappable feature.

Note that the verbal modality is crucial to activate the concept 'saliva' – surely viewers would not associate bubbling water, or an underwater seascape generally, with saliva, or a mouth. This would thus be an example of what Barthes (1986) calls text 'relaying' visual information: the verbal modality provides salient information not present in the visual modality and vice versa; hence they complement each other (for more detailed suggestions of how verbal and visual information may be related, see Unsworth and Clérigh, 2009; Bateman, 2008). The soundtrack suggests bubbling water, and perhaps swallowing, while the music is of the high-tone and eerie kind regularly used to accompany underwater footage in fiction and documentaries – except for the brief scene in which the affected coral is shown, when the tones are lower and there is an ominous 'whoosh' sound.

Shell *V-Power* petrol (commercial)

Beautifully coloured fish (reminiscent of the co-temporary Disney success *Finding Nemo*, Andrew Stanton, USA, 2003) are seen swimming in the sea in this commercial, shown on Dutch TV. A school of fish all of a sudden mechanically stops swimming at exactly the same time. A small fish narrowly escapes from a swordfish by quickly dashing away. On the sound track we hear traffic sounds – revving motors, screeching tyres, car horns, a siren. The voice-over reinforces the references to the traffic domain, and promotes Shell's *V-Power* as guaranteeing cars' optimal performance (the commercial can be accessed via Forceville, 2007). The traffic domain is thus cued by both non-verbal sound and spoken language. In turn, these steer interpretations on the visual level, such as that the suddenly motionless fish 'stop for a traffic light' and a small fish tries to escape the 'police car' swordfish. Clearly, then, the two input spaces are CARS and FISH. While first-time viewers are bound to be puzzled by the opening, once they realise that this is a commercial for petrol they will realise that the metaphor is CARS ARE FISH, not FISH ARE CARS, since petrol is metonymically related to cars rather than to fish. In the final part, we see a number of shells lying on the bottom of the sea transform into Shell logos. The similarity between the two domains is thus created via the brand name 'Shell'. Shells trigger the sea domain, which subsequently make the presence of fish a 'normal' matter.

We can infer from the voice-over – which nowhere refers to the FISH domain – that the promotable attribute is that petrol enables better performance. But the visual realm of a paradise-like underwater seascape strongly connotes 'unpolluted nature', and no doubt Shell hopes that viewers map the attribute of 'contributing to an unpolluted environment' on to the product. The showing of 'shells' constitutes the 'naturalisation' (see Barthes, 1986: 34) of the scenario of CARS-AS-FISH. The explicit performance claim of the product is thus made explicitly via the verbal realm, while the suggestion that Shell contributes to a cleaner environment is weakly implicated. And the more weakly implicated a message is, the more its derivation is the responsibility of the message's receiver rather than of its sender (see Sperber and Wilson, 1995: 199; Forceville, 1996: 93, 2005, 2009). The choice of modalities to help create similarity between the two domains and to trigger relevant mappings is thus well-considered, and is a crucial aspect of *form* that helps make the bisociation creative.

Conclusion

Summarising on the basis of the five case studies discussed, the procedure in which pertinent product attributes are saliently communicated using metaphors, the following picture arises. From a production perspective, first of all these attributes are to be decided on. In marketing terminology, this used to be called the Unique Selling Proposition (USP), the dimension in which the product advertised supposedly differs markedly from its competitors (Brierley, 1995: 140–41). But nowadays, this USP may be the creation of a specific 'brand personality' (see Koller, 2009), or a specific emotion or atmosphere as much as an attribute of the product itself. Subsequently, the advertising agency's creative team needs to find a semantic domain/mental space that can serve as the source domain from which the desired attributes can be generated, ready to be mapped in a metaphor of which the product is the target. The next step is to create similarity between the (domain of the) product and the source domain. In Blending Theory terms, this means emphasising the generic space. This similarity may be latent-but-pre-existent, but often is *created* – since this similarity usually is no more than the excuse, or quasi-motivation, for presenting the metaphor. Thus there is no pre-existent salient similarity between this particular orang-utan's and Mona Lisa's smile, between the coffee machine and a motorbike, between toothpaste containers and fish, or between car and fish. The Nespresso coffee machines are somewhat exceptional: inasmuch as the cityscapes have been the inspiration for the design of the machines, the similarity is not coincidental: the metaphor is deployed to *reinforce* a pre-existent similarity.

Finally, the choice of mode/modality is crucial for the success of the metaphorical bisociation. The chosen modality/modalities may (help) (a) identify the source domain; (b) create similarity between target (usually: the product) and source; and/or (c) cue the features to be mapped from source to target. It is thus not only the conceptual *content* of the feature(s) to be mapped on to the product that co-determine success, but also *how* these features are communicated. The fact that the similarity is to a considerable extent *created* is an important reminder that, although it may be correct to claim that many (most?) creative metaphors can be reformulated as manifestations of conceptual ones (Lakoff and Turner, 1989: 26), this approach underestimates the importance that *formal* resemblance between target and source (the catalogue of opportunities differs per medium, but can exemplify similarity of sound, form, colour, texture, pitch, framing, movement . . .) can play to make a metaphor successful. The question of the manner in which the different tasks – identifying the source, creating similarity between target and source, and cueing mappable features – are distributed over the various modalities can only be answered by systematic corpus research. Such corpus research needs to take into account factors including medium (static versus moving images), product category, target audience, and possibly (sub)cultural variation. Ideally, the findings are subsequently tested in experiments in which the salient factors are systematically varied. In the cases discussed above it can be noted that all the commercials anchor visually or sonically cued mappable features by means of language, whether in superimposed text, voice-over, or both. In the Artis billboard, whereas the verbal modality is crucial for the identification of the metaphor in the first place, only the word 'live' can be seen as hinting at a mappable feature ('come to Artis to see a *living* Mona Lisa rather than go to the Louvre to see a mere painting of a long dead one!'). In the Nespresso ads, the mappable feature is admittedly presented verbally ('high design for Nespresso') but, due to its small print, with little salience. Presumably many readers flipping through the magazine will not read this line, nor need to do so in order to get the message.

I speculate that favouring non-verbal modes in the creation of multimodal advertising metaphors has the following advantages over a restriction to the verbal mode alone: sound, music, and visuals

(i) communicate on a more directly emotional level than words do;

(ii) require the addressee to construe the metaphor and/or its mappable features him/herself, resulting in more cognitive effort on the part of the addressee and hence in more attention to the commercial message;

(iii) lacking the metaphorical 'is' of language, enable the creation of metaphors that would have been improbable or ridiculous if rendered in purely verbal terms;

(iv) steer addressees into making inferences at their own responsibility and discretion, allowing for a degree of 'personalisation' of a mass-medium message.

References

Barthes, R. (1986/1964) 'Rhetoric of the image' in *The responsibility of forms*, Barthes, R. (ed.) (trans. Howard, R.). Oxford: Blackwell, 21–40.

Bateman, J. (2008) *Multimodality and genre: A foundation for the systematic analysis of multimodal documents*. Houndmills: Palgrave Macmillan.

Black, M. (1979) 'More about metaphor' in *Metaphor and thought*, Ortony, A. (ed.) Cambridge: Cambridge University Press, 19–43.

Boden, M. A. (2004) *The creative mind: Myths and mechanisms* (2nd edn). London: Routledge.

Brierley, S. (1995) *The advertising handbook*. London: Routledge.

Câmara Pereira, F. (2007) *Creativity and artificial intelligence: A conceptual blending approach*. Berlin: Mouton de Gruyter.

Fauconnier, G. and Turner, M. (1998) 'Conceptual integration networks'. *Cognitive Science* **22** 133–87.

Fauconnier, G. and Turner, M. (2000) 'Compression and global insight'. *Cognitive Linguistics* **11** 283–304.

Fauconnier, G. and Turner, M., (2002) *The way we think: Conceptual blending and the mind's hidden complexities*. New York: Basic Books.

Forceville, C., (1995) 'IBM IS A TUNING FORK: Degrees of freedom in the interpretation of pictorial metaphors'. *Poetics* **23** 189–218.

Forceville, C. (1996) *Pictorial metaphor in advertising*. London: Routledge.

Forceville, C. (2004) 'Review of Fauconnier and Turner (2002)'. *Metaphor and Symbol* **19** 83–9.

Forceville, C. (2005) 'Addressing an audience: Time, place, and genre in Peter Van Straaten's calendar cartoons'. *Humor* **18** 247–78.

Forceville, C. (2006) 'Non-verbal and multimodal metaphor in a cognitivist framework: Agendas for research' in *Cognitive linguistics: Current applications and future perspectives*, Kristiansen, G., Achard, M., Dirven, R. and de Mendoza Ibáñez, F. R. (eds). Berlin: Mouton de Gruyter, 379–402.

Forceville, C. (2007) 'Multimodal metaphor in ten Dutch TV commercials'. *Public Journal of Semiotics* **1** (1) 19–51. http://www.semiotics.ca/

Forceville, C. (2008) 'Metaphor in pictures and multimodal representations' in *The Cambridge handbook of metaphor and thought*, Gibbs, R. W. Jr (ed.). Cambridge: Cambridge University Press, 462–82.

Forceville, C. (2009) 'Relevanz und Prägnanz: Kunst als Kommunikation' (trans. Plümacher, M.). *Zeitschrift für Semiotik* **31**(1–2) 31–63.

Forceville, C., Veale, T., and Feyaerts, K. (2010) 'Balloonics: The Visuals of Balloons in Comics' in *The Rise and Reason of Comics and Graphic Literature: Critical Essays on the Form*, Goggin, J. and Hassler-Forest, D. (eds), Jefferson, NC: McFarland, 56–73.

Forceville, C. (2011) 'Practical cues for helping develop image and multimodal discourse scholarship' in *Bilder, Sehen, Denken: ZumVerhältnis von begrifflich-philosophischen und empirisch-psychologischen Ansätzen in der bildwissenschaftlichen Forschung*, Sachs-Hombach, K. and Totzke, R. (eds). Cologne: Von Halem, 33–51.

Forceville, C. (in prep.) 'Creative visual variation in comics balloons.'

Forceville, C. and Urios-Aparisi, E. (eds) (2009) *Multimodal Metaphor*. Berlin: Mouton de Gruyter.

Gibbs, R. W., Jr, (ed.) (2008) *The Cambridge handbook of metaphor and thought*. Cambridge: Cambridge University Press.

Gineste, M.-D., Indurkhya, B. and Scart, V. (2000) 'Emergence of features in metaphor comprehension'. *Metaphor and Symbol* **15** 117–35.

Grady, J., Oakley, T., and Coulson, S. (1999) 'Blending and metaphor' in *Metaphor in cognitive linguistics*, Gibbs, R. W. Jr and Steen, G. J. (eds). Amsterdam: John Benjamins, 101–24.

Holyoak, K. J., Gentner, D. and Kokinov, B. N. (2001) 'Introduction: The place of analogy in cognition' in *The analogical mind: Perspectives from cognitive science*, Gentner, D., Holyoak, K. J. and Kokinov, B. N. (eds). Cambridge: MIT Press, 1–19.

Jewitt, C. (ed.) (2009) *The Routledge Handbook of Multimodal Analysis*. London: Routledge.

Koestler, A. (1969 [1964]) *The act of creation* (with a new preface by the author), The Danube Edition. London: Hutchinson.

Koller, V. (2009) 'Brand images: Multimodal metaphor in corporate branding messages' in *Multimodal metaphor*, Forceville, C. and Urios-Aparisi, E. (eds). Berlin: Mouton de Gruyter, 45–71.

Kress, G. (2009) 'What is mode?' in *The Routledge handbook of multimodality*, Jewitt, C. (ed.). London: Routledge, 54–67.

Kress, G. (2010) *Multimodality: A Social Semiotic Approach to Contemporary Communication*. London: Routledge.

Kress, G. and Van Leeuwen, T. (2001) *Multimodal Discourse: The Modes and Media of Contemporary Discourse*. London: Arnold.

Lakoff, G. and Johnson, M. (1980) *Metaphors we live by*. Chicago: University of Chicago Press.

Lakoff, G. and Johnson, M. (1999) *Philosophy in the flesh: The embodied mind and its challenge to Western thought*. New York: Basic Books.

Lakoff, G. and Turner, M. (1989) *More than cool reason: A field guide to poetic metaphor*. Chicago: University of Chicago Press.

Sperber, D. and Wilson, D. (1995) *Relevance theory: Communication and cognition* (2nd edn). Oxford: Blackwell.

Turner, M. and Fauconnier, G. (1995) 'Conceptual integration and formal expression'. *Metaphor and Symbolic Activity* **10** 183–204.

Unsworth, L. and Clérigh, C. (2009) 'Multimodality and reading: The construction of meaning through image-text interaction' in *The Routledge handbook of multimodality*, Jewitt, C. (ed.). London: Routledge, 151–63.

Veale, T., Feyaerts, K. and Forceville, C. (in prep.) *Hiding in plain sight: The art of creative duality in words and images*.

Acknowledgements

The seeds of this chapter were planted when the author collaborated with Kurt Feyaerts and Tony Veale on the project *The Agile Mind: Creativity in Discourse and Art*, at the Royal Flemish Academy of Belgium for Science and the Arts (VLAC), Brussels. It elaborates on one half of the presentation given at the *Discourse and Creativity* conference at Hong Kong City University (May, 2009). The other half, on creativity in comics balloons, will appear elsewhere (Forceville, in prep.; see also Forceville *et al.*, 2010). While I have discussed several advertisements and commercials examined here in earlier papers and chapters, the focus on their creativity is new. I thank Rodney Jones for his editorial suggestions on an earlier draft.

CHAPTER 8

Design, production and creativity

Theo van Leeuwen

Semiotics and creativity

The project of social semiotics, as I see it, involves three interrelated practices: the study of semiotic resources, the study of semiotic practices, and the study of semiotic creativity, that is, of the way new semiotic resources and practices and new uses of existing semiotic resources are invented.

The study of semiotic resources originated in the study of language but has more recently been extended into the study of other semiotic resources – into the forms and meanings of bodily expression and self-presentation, and into art and music and their applications in the design of our everyday world. I have become increasingly convinced that semiotics has to be both systematic and descriptive, to understand how physical materials and actions are organised into semiotic systems, and a cultural-historical study, to understand why they are organised the way they are. Without this it is too easily forgotten that semiotic resources are organised by human actors in the context of the needs and exigencies of human social practices.

The study of semiotic practices, too, has to be both descriptive and sociological. It needs to describe, precisely and in detail, how people produce and use semiotic resources, but it also needs to explain how these uses come about, how they are taught or otherwise acquired, regulated, debated, changed, and so on. And again, while the former can be done with the tools of multimodal discourse analysis, the latter needs to move beyond the text to study the processes of production and use, and the way these processes are socially represented and regulated.

These two are the types of semiotic activity I have most often studied, but increasingly I am aware that they can never be purely descriptive and explanatory. They always also intervene in the semiotic practices they describe and explain, and so contribute to social change, whether immediately or slowly and steadily. Here the link between semiotics and art is particularly important. The semiotician and the artist travel along parallel paths and they might as well talk and work together. Just one example. Gunther Kress and I studied the meaning potential of the vertical angle in images, the difference between the frontal view and the side-on view, the profile. A gap became apparent. What about the view from behind? In such a case, the semiotician collects examples of 'views from behind', tries to piece together what people have tried to express with such pictures, and then writes about it. The photographer Candice Milon asked precisely the same question, and answered it by making a series of photos which systematically anonymised people (Milon, 2010). To her, the view from behind was a resource for anonymisation, along with soft focus, hair blowing in front of the face, and other devices. The result was a fetching series of portraits, but also and at the same time a systematic analysis, albeit expressed pictorially rather than in words.

In this chapter I want to explore this third aspect of semiotics a little further and use some semiotic ideas to explore the forms of creativity that can drive semiotic change in particular cultural and historical contexts. I will draw my examples from the semiotic resources and practices of music, more specifically from jazz, but I expect that the line of thought developed here can also be applied to other semiotic modes.

The relation between semiotic resources and semiotic practices

In our book *Multimodal discourse* (2001), Gunther Kress and I devised a multi-layered model of semiotic production and interpretation. We recognised four layers: discourse, design, production and distribution, and used them to relate two things to each other: the semiotic activities of understanding, designing, producing, and 'distributing', and the different kinds of semiotic resources that are used in each of these activities.

'Discourse' we interpreted in a Foucauldian sense, as a construction of some aspect of reality that then constitutes a framework for understanding that aspect of reality which can be drawn upon when that aspect of reality is to be represented in some way or other. The same discourse can be drawn upon in different kinds of text. To give an example, in the 1960s a new African-American discourse about Africa emerged that found expression in words, for instance in black histories, as well as in art and media, for instance in the television series *Roots*, and also in the music of key jazz musicians of the period.

'Designs' we described as formats or templates for embedding discourses in specific social contexts. What they add to the discourse is, above all, a communicative purpose and a particular relation between those who participate in the communication, for instance between the partners in a conversation, or between performers and their audiences.

In music the blues is such a design schema. Its traditional communicative purpose is to convey a usually stoic or world-weary piece of wisdom about life, which is first stated, then reiterated, then in some way elaborated or challenged, and finally repeated again, as here in the Billie Holiday blues *Fine and Mellow*:

> Love is just like a faucet   It turns off and on;
> Love is just like a faucet   It turns off and on;
> Sometimes when you think it's on baby,   It has turned off and gone.

The music has a similar structure. It states, and then repeats a melodic line, perhaps one that goes down in pitch, then 'challenges' that line through a harmonic and melodic change (e.g. a line in which the pitch goes up), to then return to the original musical statement. But the musical schema also encodes a relation between the performer and the audience. There is a gap after each line, to be filled with audience reactions ('Oh yeah', 'Yeah baby' and so on), in call and response fashion. In more stylised versions the response may be played by one or more of the accompanying instruments.

Four more things can be said about these design schemas. First, they exist at various levels. There are design schemas for different kinds of call and response 'adjacency pairs' (cf. van Leeuwen, 1999: 71–7) and design schemas for whole verses or choruses, just as, in language, there are design schemata for clauses and design schemas for texts (e.g. narrative schemas). Secondly, specific contexts prescribe what design schemas should be used, and do so with different degrees of detail and strictness. Old style country blues, for instance, had different numbers of bars, for example 8 or 10 or 14, but later the format became standardised to 12 bars. The design schemas of classical music, on the other hand, do not prescribe specific amounts of bars, but they do prescribe exactly what notes should be played and for how long, while blues performers are free to embellish and vary their phrasing.

Thirdly, a given design schema can be realised in different communicative modes – a clause can be spoken or written; a narrative can be written or filmed or in the form of a comic strip; a blues can be instrumental or vocal; and so on. And even though there exist norms that prescribe which design schemas go with which modes of communication, breaking such norms is often a powerful source for semiotic innovation (e.g. constructing a film on the basis of a musical schema). Finally, design schemas add their own meanings. They are not a neutral vehicle for discourses. The difference between monologic and dialogic design schemas, for instance, invokes different social relations, as well as the values and sentiments these involve. Concepts such as 'participation', 'interaction', etc. are value-laden, and so are the design schemas that actualise these concepts.

'Production' is the process which gives material form to the design, and the qualities of the materials and actions that perform this work add further meaning. The voice of Billie Holiday, for instance, is as important a part of the message as the words and notes she sings, as will be discussed in more detail below.

'Distribution', finally, is the term we use for the technical re-coding of a text or communicative event for purposes of distribution and/or preservation, for instance the record or CD on which a performance by Billie Holiday is preserved. 'Distribution' media add meanings of their own. A film is not the same when seen on a small screen or a big screen, for instance, and recording and mixing techniques can not only manipulate the volume of different voices or instruments, but also alter their quality. As a result the dividing line between the performer and the technician is, today, increasingly blurred.

As illustrated in Figure 8.1, our four layers have been inspired both by theories of stratification (Hjelmslev, 1961) and by Erving Goffman's theory of footing (1981), and in fact try to bring the two together – the 'layers' or 'strata' of semiotic resources, on the one hand, and the roles involved in semiotic practices on the other hand. 'Discourse' is the semiotic resource

Kress & van Leeuwen	Hjelmslev	Goffman
Discourse	Content	Principal
Design	Expression	Author
Production	Substance	Animator
Distribution		('loudspeaker or telephone')

FIGURE 8.1 The relation between semiotic resources and semiotic practice

used by the 'principal', the person or institution 'whose beliefs are told', 'whose position is established' (Goffman, 1981: 144). The Billie Holiday blues, for instance, draws on a discourse of romantic love in which love is beyond our control, affecting us like a disease. This discourse is not of Billie Holiday's making, but a cultural creation which goes back to the courtly cultures of the late Middle Ages, and has since informed countless songs, poems and stories, as well as the beliefs and lives of countless people.

Design is the semiotic resource used by what Goffman calls the 'author', the person 'who selects the sentiments that are being expressed and the words in which they are encoded' (ibid.). Billie Holiday, for instance, uses the design schema of the blues, thus embedding the discourse of romantic love in a particular mode of communication with a particular communicative purpose. 'Production' is the semiotic resource used by what Goffman calls the 'animator', the 'sounding box in use' (ibid.), in our example Billie Holiday's voice – but another 'animator' might perform the same 'design' in a very different way. What is not always noticed, finally, is that Goffman also refers to a role that corresponds to our fourth category, when he says that 'animators' 'sometimes share this physical function with a loudspeaker system, or telephone' (ibid.). In modern semiotics this function will have to become far more important than it has been so far, as 'production' and 'distribution', while theoretically still distinct, become more and more intertwined, so that 'producers' have to become more technically adept, and 'technicians' more creative.

In what follows I will focus on 'design', 'production' and 'distribution' as distinct creative catalysts. I will discuss, in turn, the creation of new design resources and practices; the creation of new production resources and practices; and the creation of technical resources that rewrite the traditional distinction between 'production' and 'distribution'.

I should mention, however, that the layering I have described in this section applies equally to the interpretation and appreciation of semiotic artifacts and events. The interpreter, too, needs to draw on all four kinds of resources – discourses, design schemata, production experiences and awareness of the relevant technological factors. Which resources will be invoked will depend on the regimes of interpretation and appreciation that dominate in the given context. The same film will be interpreted differently in a university film studies essay and in a conversation with a group of friends, after watching the film on TV or in a cinema. And just as there is creativity in producing semiotic artifacts and events, so there is also creativity in thinking up new resources for interpretation and appreciation. This kind of creativity is in fact precisely the kind of contribution semioticians enjoy making.

Design and creativity

The dominant design schema in jazz has always been the cycle of harmonic transitions, the 'chord schema', often taken from the chorus of Broadway songs, and typically with 36 bars per cycle. Most jazz performances are therefore based on designs that are not created by the performers. Yet, just as a limited number of narrative schemas can give rise to endlessly varied stories, so these schemas can generate endlessly different performances, and in jazz it is in fact a mark of distinction to be able to blow new life into familiar tunes such as 'Body and Soul' or 'The Girl from Ipanema' – or, indeed, the blues schema.

But, perhaps taking a leaf out of the book of the twentieth-century European composers who sought to remake the language of music, modern jazz musicians have been keen to create new design schemas. John Coltrane, somewhere in the early 1960s, wondered why the chord progressions of so many songs were so similar, endlessly recycling the 2-5-7 progression, as for instance in songs like *Autumn Leaves*. So he devised an entirely novel and highly unusual chord progression, with jumps of a fifth between subsequent chords, wrote a composition based on this schema, and appropriately called it 'Giant Steps'. It turned out to be difficult to perform with. Coltrane could improvise magnificently on it, but the otherwise fine pianist Tommy Flanagan, on the first recording of 'Giant Steps', floundered. More than 45 years later, 'Giant Steps' is still a challenge to most jazz musicians. Anthony Braxton used another design schema, taking the rhythmic pattern of melodies, rather than their underlying chords, as the basis for improvisation. This brought remarkable coherence to atonal improvisation. Sometimes such ideas are worked out in compositions, sometimes they develop in the course of the performer's practice, but they always added to the language of jazz.

This approach to creativity characterised not just 'modern jazz', but all modern art. In literature, 'genre literature' was looked down on. Each work had to create its own, new design schema. Painters invented the design schemas of Cubism, de Stijl and so on, which later became part of the language of design, applied to fabrics, wrapping paper, interior design, etc. Composers rewrote the language of music, inventing atonal music, serial music, minimalist music. All of these were new design schemas, new forms of, and for, representation.

In Postmodernism the focus on creating new semiotic resources was replaced by a focus on creating new uses of existing semiotic resources, especially by combining existing semiotic resources in new ways ('fusion', in music). This also played a role in the 'localisation' of globally produced and disseminated music, the combining of global genres and local traditions (cf. Mitchell, 1996). Creation was replaced by choice. This widened the range of options in any given context, but halted the development of new design schemas, and replaced it with the circulation of existing ideas, in numerous and ever new combinations and via numerous and constantly changing media and combinations of media.

In my own musical practice of that time, I began to combine the melodies and tonalities of sixteenth-century Genevan psalms with the kind of repetitive rhythms John Coltrane had developed in his work of the early 1960s. As a minister's son in the Netherlands, I had grown up with church music, then discovered jazz, in my early teens, listening to *The Voice of America* which broadcast jazz for American soldiers in Germany. Music allowed me to bring together the two conflicting discourses expressed in these musics, and the two conflicting identity models they embodied, the one religious, the other worldly, the one slow and solemn, the other rhythmic, the one steeped in a particular Dutch religious tradition, the other coming from the USA after World War 2. The titles of the compositions (e.g. 'Rebuke Me Not') sometimes raised eyebrows in jazz clubs.

Production and creativity

As I have explained in earlier work (e.g. van Leeuwen, 2009), particular sound qualities can create experiential metaphors, allowing us to understand complex and abstract ideas on the basis of concrete experiences. These experiences are of two kinds: experience of how we produce sounds with our vocal apparatus and what the resulting sounds sound like, and experience of where and why we might produce them.

Such experiences are a major resource for creating new meaning with sound, and new understandings of sounds, especially in cases where we cannot understand sounds on the basis of established cultural references.

Take, for instance, 'rough' sounds. Our experience tells us where roughness in the voice comes from – wear and tear, a hard life, smoking and drinking. And we can transfer that knowledge to hear in a sound the *idea* of roughness or harshness, and to link that idea to other things that are expressed at the same time. As I have written elsewhere (1999: 132):

> The meaning of roughness lies in what it is: rough. The rough voice is the vocal equivalent of the weatherbeaten face, the roughly plastered wall, the faded jeans, the battered leather jacket.

Or take vocal tension. Here, too, we know what it sounds like – higher, sharper and brighter than the more relaxed voice and what causes us to produce it – fear, anticipation, excitement. In specific contexts this can then create more specific meanings. The ethnomusicologist Lomax (1968: 194) wrote of the tension in the preferred female singing voices in societies with severe sexual repression:

> It is as if one of the assignments of the favoured singer is to act out the level of sexual tension which the customs of the society establish as normal. The content of this message may be painful and anxiety-producing, but the effect on the culture member may be stimulating, erotic and pleasurable since the song reminds him of familiar sexual emotions and experiences.

So vocal qualities are universal, common experiences. But their uses and the values and sentiments that are attached to them, are the product of specific cultural histories.

The great singers and actors of the twentieth century created new, and instantly recognisable voices by combining such qualities in specific ways. Thus Billie Holiday's voice is at once relatively high, which makes her sound vulnerable, as women tend to use lower voices when they want to be dominant and assertive, and relatively loud, giving the impression that she does want to be heard, does want to assert herself. It is also a tense voice, suggesting pain and anxiety, with a bit of a rough edge, and just a touch of vibrato on the sustained notes. Her voice thus combines vulnerability, tension and toughness, and the restraint as well as the expression of emotion (again, our experience tells us that vibrato, the trembling of the voice, is a sign of emotion). It is this combination that makes her voice unique and touching.

Until the twentieth century, the voices of actors, singers and public speakers had been disciplined towards uniform and institutionalised standards of propriety and aesthetic ideals. Vocal 'production' had been quite prescriptive, even though this never led to complete uniformity. Amplification changed this, opening up new timbral resources and allowing singers and actors to create new voices to express new identities, and, in the process, a new language of the voice.

In Postmodernity this, too, changed. As Michel Chion (1999: 173) notes of American screen actors, they no longer project a single, instantly identifiable persona with their voice, but adjust their voice quality to the role or scene they are playing:

Compare two roughly contemporaneous Dustin Hoffman movies. In Barry Levinson's *Rain Man*, he has a metallic and nasal voice and in Stephen Frears' *Hero* it is coarser. If you listen to both films without the picture, it is quite difficult to identify both voices as coming from the same actor.

Marlon Brando's hoarse whisper is now part of the repertoire of many American actors, a resource to employ whenever a sense of brooding, yet sensuously attractive menace is required. Singers like Madonna or Gwen Stefani, to just mention two examples, also draw on a wide range of recognisable singing styles. The Gwen Stefani song 'Bubble Pop Electric', for instance, evokes a 1950s teenage date: 'Drive in movie. Drive in move me'. The setting, that era of 'penny loafers and poodle skirts', and the impatient anticipation of 'petting' in the back seat of a car are portrayed, not only by the lyrics, but also by the timbre of Gwen Stefani's voice, which alternates between seduction, as she addresses her boyfriend in a low, breathy voice ('Tonight I'm falling, won't you catch me'), and teenage pop culture, as she repeats the words 'Bubble Pop Electric' in a higher, tenser voice with a somewhat artificial ('electric') sheen. Just as her music combines many different styles, so her voice has many different registers, allowing her to express complex feelings and complex identities.

In Postmodernity the impetus to create new semiotic resources, new timbres in this case, stopped. The repertoire stabilised. Voices began to be understood, not on the basis of experience, but on the basis of cultural connotation, of 'where they come from'. Marlon Brando's voice, for instance, would now be understood holistically, and on the basis of familiarity with his most famous roles, rather than more directly, on the basis of our understanding of its timbral qualities.

Distribution and creativity

On a modern digital piano I cannot do what Billie Holiday did with her voice, or what I can even do, to some extent, on an acoustic piano. I cannot make the sound rougher or tenser as I play. Instead the designers of the instrument have provided me with a choice from a wide range of piano sounds, some tenser, some mellower, some higher, some lower, some smoother, some rougher (in a synthetic kind of way) and so on. All I can do myself is phrasing.

An earlier instrument of which the same could be said was, and still is, the church organ. Its voice was the voice of God, not that of the organist.

Until the eighteenth century it was not generally accepted that anyone but God could create (Williams, 1985: 83). *'Creatura non potest creare'*, St Augustine said, 'Creatures cannot themselves create'. The timbral qualities in my digital piano are designed by a new kind of creator, a new kind of God, the software designer. I *can* create new sounds myself, but to do so I must change hats, become a 'technician' rather than a musician. I cannot do it as a player, while I play.

Software designers are now the key creators, refashioning and recombining our key semiotic resources for us, and providing us with more choice, yet less ability for truly individual variation at the 'production' level than we have ever had. And that does not just apply to musical instruments, but also to language, at least written language. Language design and language production, too, are increasingly regulated, not by school teachers, editors, style manuals and so on, but by the options and procedures built into the software we use. Yet it is always possible to reconnect with the physicality that ultimately underlies all semiotic production and to explore the many new semiotic resources that wait to be discovered, and the many old ones that wait to be rediscovered.

References

Chion, M. (1999) *The voice in cinema*. New York: Columbia University Press.

Goffman, E. (1981) *Forms of talk*. Philadelphia: Philadelphia University Press.

Hjelmslev, L. (1961) *Prolegomena to a theory of language*. Madison, Wisconsin: University of Wisconsin Press.

Kress, G. and Van Leeuwen, T. (2001) *Multimodal discourse – The modes and media of contemporary communication*. London: Arnold.

Lomax, A. (1968) *Folk song style and culture*. New Brunswick, New Jersey: Transaction Books.

Milon, C. (2010) http://www.milim.com/work.php@id=148 (accessed 22 September 2010).

Mitchell, T. (1996) *Popular music and local identity – Rock, pop and rap in Europe and Oceania*. London: Leicester University Press.

Van Leeuwen, T. (1999) *Speech, music, sound*. London: Macmillan.

Van Leeuwen, T. (2009) 'Parametric systems: The case of voice quality' in *The Routledge handbook of multimodal analysis*, Jewitt, C. (ed.). London: Routledge, 68–78.

Williams, R. (1985) *Keywords*. London: Fontana.

CHAPTER 9

The creation of a community artist in everyday life: Long-duration process and creative actions

Sigrid Norris

> One weekend, not long after her then-husband had moved out, and a weekend, which the children spent with their father, Andrea suddenly felt the need to paint. She drove to an art store, bought paint and brushes, went back to her house, turned the music on really loudly, and began to paint on walls.
>
> (Norris, 2011: s239)

Introduction

I have conducted a good number of ethnographic studies of social actors acting in their everyday life in the last ten years and have found that many social actors paint, write poetry, or dance, but only a few of them are considered community artists, poets, or dancers. How is it, I ask in this chapter, that some actions are actions that produce *one* social actor as a community artist while the same kinds of actions may not produce *another* social actor as a community artist. Here, I take a close look at one woman, Andrea, who at the time of study, began to paint. The data pieces come from a one-year ethnographic study I conducted in Germany in 1999–2000. Andrea, soon after having painted her first painting, was viewed by her immediate network as an artist. An obvious answer would be that there was a strong quality of artistic expression; however, here, I will stay away from evaluating the art work as such, but will investigate this case in a more holistic way, illustrating how the identity production of *community artist* has come about.

When I asked friends and members of Andrea's network how some actions can produce one social actor as an artist, but not another, they claimed that the actions that produce one social actor as an artist are *creative*, while the actions of the other are *not-really-creative*. When probing further about what is and what is not considered creative, these social actors in Andrea's network would often say similar things to Helga in the following excerpt.

Transcript 9.1: Helga speaks about creativity

(1) Helga: *ja klar,*
well, of course,

(2) *alle Bilder sind kreativ.*
all paintings are creative.

(3) *aber weisst du*
but you see.

(4) *da ist ein grosser Unterschied*
there is a big difference

(5) *wenn ich male*
if I paint

(6) *oder wenn Andrea malt.*
or if Andrea paints.

(7) *meins ist schon kreativ*
mine is (emphasis) creative

(8) *so für mich mein ich,*
so for myself I mean,

(9) *für mich,*
for me,

(10) *sicher,*
surely,

(11) *wenn ich es male.*
when I paint it.

(12) *aber davon*
but because of that

(13)	*bin ich noch lange kein Künstler.* I'm still not an artist.
(14)	*da muss man schon* for that you have
(15)	*richtig kreativ sein* to be really creative

Because of this quite common opinion of social actors when referring to creative actions, I will use the term *creative* creative actions and the term *not-really-creative* creative actions to refer to those actions that fall into the categories summarised in Table 9.1.

The term creative actions is thus (for the analysis in this chapter) limited to actions that are product creating and product focused and simultaneously are a form of self-expression and can either be produced for the self or for others. Both *creative* and *not-really-creative* creative actions thus do not differ in these respects. Yet, social actors in everyday life claim that these actions do differ, usually by alluding to the quality embedded in the artwork produced, calling it either *creative* or *not-really-creative*. However, as I will illustrate with the help of Andrea's very first painting that she produced, what was viewed in her network as *creative* (and what was not) was in fact not so much located in the product itself; and, in fact, had indeed not much to do with the quality of artistic expression, but rather was located in a long-duration process including many different actions that led to the art work's production in the first place and the material used in the second place. In other words, *creativity* was located in the combination of mediational means used over time, the theme and materiality for the first painting chosen, and the situatedness of the painting conveying emotive meaning. While I allude to the existence of *creative* and *not-really-creative* creative actions as the members of Andrea's network distinguish them and speak about them, my true focus in this chapter is the production of a *community artist* whereby

TABLE 9.1 *Creative* and *not-really-creative* creative actions as viewed by social actors in everyday life

	Creative creative-actions	*Not-really-creative* creative-actions
Product creating	X	X
Product focused	X	X
A form of self-expression	X	X
Produced for the self	X either	X either
Produced for others	X or	X or

I illustrate that what social actors view as an outcome of *creative* actions in fact is to be found in a long-duration process.

Methodology

I develop my argument by returning to my one-year ethnographic study of Andrea. During this ethnography, Andrea began to paint on the particular weekend referred to in the extract at the beginning of this chapter and developed the beginning of an artistic career. Now, ten years later, Andrea is a well-respected artist in her community and the owner of an art school where she teaches adults and children to paint. In this chapter, I only look at the art form of painting in particular, and while some aspects of this example of one artist may be generalisable, I will largely stay away from generalising in this chapter.

When Andrea began to paint on the walls of her house, she was immediately considered to be an artist by members in her network. Andrea began her artistic career – just as many social actors who paint, do – by copying and elaborating a piece of art that she admired. This led me to the question of why some creative actions are viewed as actions that produce art while others do not. I believe one possible answer can be found in a long-duration process rather than the action itself. In order to illustrate this point, I look closely at mediated actions (Wertsch, 1998; Scollon, 1998, 2001) over long-duration time scales (Lemke, 2000) that produce a developing process.

In order to shed light upon the creative mediated actions *per se* as well as upon the development of the creative process involved, I use multimodal interaction analysis (Norris, 2004, 2011) as my methodological framework, examining the lower-level, higher-level and frozen actions which my participant (co)produces together with other social actors and/or the environment.

A mediated action is defined as a social actor acting with or through a mediational means (Wertsch, 1998). A mediational means for example would be a brush but also the hand that uses the brush and the arm that moves it across a wall or a canvas. A lower-level action would be a particular brush stroke, a higher-level action the action of painting a piece of art, and the frozen action an embedded brush stroke (on a lower level) and the action of a particular artist having painted the piece embedded on the canvas or the wall in the above example (on a higher level). In other words, the artist's lower-level actions and the use of the mediational means such as brush, colour or movement are frozen in the finished art work. Similarly, the artist's higher-level action of painting as observable in the style, the use of colour, and the development of imagery is frozen in the art work and can at a later point in time be traced back to that particular social actor.

Mediational means, as discussed by Wertsch (1998) and Scollon (1998, 2001), are always multiple and they form an important part in the actions performed. An action, in fact, cannot be comprehended without our considering the social actor performing the action and the multitude of mediational means involved.

Actions

Since every action in multimodal interaction analysis (Norris, 2004, 2010) following mediated discourse analysis (Scollon, 1998, 2001; Norris and Jones, 2005) is considered to be a mediated action, we always have to consider the components that make up an action:

1 the social actor(s);

2 multiple mediational means; and

3 the performance of the action that ensues.

The performance of an action can thus never be understood without the consideration of *who* the social actors are that perform the action, or without the consideration of the multiple mediational means involved *to* perform the action. Actions can also be viewed in objects in a frozen form (Norris, 2004). Objects, which are mediational means as well, entail actions such as the fact that they have been bought by a social actor, that they have been placed in a particular space by a social actor and so on.

Analysis

A creative action is a mediated action, which is linked to a particular social actor, to particular mediational means, and to particular actions that ensue and to those that precede it. In my example, the social actor is Andrea; the mediational means that I focus on are narratives, books, paint, brushes, walls, and Andrea's *Scream* shown in Figure 9.1 (p. 154); and the actions that I focus on are Andrea's telling of a narrative, collecting books, and painting on a wall.

The social actor: Andrea

Andrea was going through a divorce when she began to paint. While she had not been painting before, Andrea had a long history in which she told and retold the following story, which was told to the researcher a few weeks before Andrea began to paint.

Transcript 9.2: Andrea had wanted to study art

(1)	Researcher:	*und* and
(2)		*was hast du studiert* what did you study
(3)	Andrea:	*ja nich das* well not what
(4)		*was ich eigentlich wollte.* I had wanted to.
(5)	Researcher:	*was wolltest du denn studieren* what did you want to study
(6)	Andrea:	*ich habe Architektur studiert* I studied architecture
(7)		*aber eigentlich* but actually
(8)		*wollte ich Kunst studieren* I wanted to study art
(9)		*das wollten meine Eltern aber nicht.* but my parents did not want me to.
(10)		*haben sie mir ausgeredet* they talked me out of it
(11)		*von Kunst kann man nicht leben,* you can't live from art,
(12)		*dann steh ich am Ende auf der Strasse,* then in the end I'd be living on the street,
(13)		*und so* and such
(14)		*nach dem Motto* kind of like that
(15)		*und du glaubst gar nicht,* and you wouldn't believe,
(16)		*wie oft ich das schon bereut habe.* just how often I have regretted that.

A few weeks later, Andrea had performed the action of painting on the walls of her house and friends came over to look at her art work. She had painted the *Scream* as her very first painting. Helga, who was one of the first to view it, said the following:

Transcript 9.3: Helga views Andrea's wall-painting

(1)	Helga:	*toll!* great!
(2)		*das find ich jetzt echt toll!* I think this is absolutely great!
(3)		*man sieht das super aus.* that really looks fantastic.
(4)		*so richtig Kunst.* real art.
(5)	Andrea:	*meinst du?* you think?
(6)	Helga:	*ja.* yes.
(7)		*der Künstler steckt eben in dir* the artist is inside of you
(8)		hhh
(9)	Andrea:	hhh
(10)		*hab ich ja immer gewusst,* I've always known this,
(11)		hhh
(12)	Helga:	*nee,* no,
(13)		*aber ehrlich!* but really

Andrea was thus considered an artist from the very beginning of her painting career. But the reason why Andrea was considered an artist from the very beginning becomes clearer when we investigate the mediational means that help induce this view.

The mediational means: Narratives, books, paint, brushes, walls, and Andrea's *Scream*

Mediational means possess materiality to varying degrees. But, of course, mediational means are usually multiple at any one point. An utterance, for example, includes multiple mediational means from language itself to the pharynx, the epiglottis and the tongue and lips forming the words. As mediational means itself, an utterance possesses a very brief and fleeting materiality. It is produced and almost simultaneously heard and then has vanished. Another example would be a paint brush, which also includes multiple mediational means such as the parts that it is made of. However, a paint brush has durable materiality. It can exist over long periods of time, can be grasped, looked at, described and used.

While the material consistency between an utterance and a paint brush is certainly strong, I illustrate below that social actors use these mediational means in similar ways, producing (the mediational means usually viewed as consisting of fleeting materiality) in a way to allow them to exist over long periods of time, giving them materiality in a sense where they can also be grasped, glimpsed, described and used. I illustrate this notion by looking at what I call a narrative-skeleton.

A narrative-skeleton

As illustrated in Transcript 2 above, Andrea had always wanted to become an artist. Her narrative-skeleton about her parents talking her out of her wish to study art and into the more practical degree of architecture is a recurring mediational means for Andrea. Throughout my research, and long before the study as I could glean from accounts by social actors in Andrea's network, this narrative recurs, allowing Andrea to portray herself first as someone who wanted to be an artist; then, as someone who is a budding artist; and, finally, as someone who is an artist.

The mediational means, the narrative-skeleton (consisting of her parents and their desire that Andrea studied architecture, her own desire to study art, and her having listened to her parents), is embellished during the time of my study with the section that tells of Andrea having defied her parents' wish in the end. The narrative-skeleton is fleshed out differently when Andrea speaks with different social actors and when she tells it in different circumstances and at different times. While the narrative-skeleton thus first functioned to portray Andrea as a potential artist long before she began to paint, it is then utilised by Andrea to claim herself as an artist as she adds the newest information to it.

This mediational means, the narrative-skeleton, as fleeting as it is when being told, gains materiality in the sense that the narrative-skeleton can be grasped, glimpsed, described, and used by other social actors in her network. Thus, the sporadic recurrence of the narrative-skeleton in Andrea's own discourse and in the discourse of a great number of social actors in her network over a long period of time gives it stability. At the same time, it gives stability to Andrea's identity, first as a potential, then as a budding, and now as an accomplished artist (Norris, 2011).

Books

Andrea has always collected art books. These art books are positioned in her living room for everyone to see. Art books are also often lying open on her coffee table or on the sofa, and Andrea looks through them and shows particular pages to members in her network. Often, social actors visiting Andrea utilise a part of Andrea's narrative-skeleton or refer to it by saying something as Anna does in the following excerpt. Here, Anna is visiting as Andrea is looking at some paintings in an art book.

Transcript 9.4: Anna refers to Andrea's narrative-skeleton

(1)	Anna:	*und,* and,
(2)		*warum haste nich einfach gemacht* why didn't you just do
(3)		*was du wolltest?* what you wanted?
(4)		*hättest doch einfach* should've just
(5)		*Kunst studieren können?* studied art?
(6)	Andrea:	hm.
(7)		*aber guck mal hier* but look at this
(8)		*wie der das gemacht hat* how he did this
(9)		*echt interessant* very interesting

(10)	Anna:	*kannst du mir ruhig zeigen* you can show it to me
(11)		*aber verstehen tu ich das nich* but I sure don't understand it hhh
(12)	Andrea:	hhh

The materiality of the art books is durable. They can be grasped, looked at, described and used. When lying on a table or the sofa, these books tell of Andrea's actions of having looked through them. These are the frozen actions that tell a visitor of Andrea's interest in art, even when she does not speak about art at the time.

In Andrea's everyday life even before she began to paint, these mediational means, the narrative-skeleton with its stabilising effect over long periods and the art books with their consistent durable materiality and their embedded frozen actions, help to continuously point to Andrea first as someone who wants to know about art and second as someone who knows more about art than other social actors in her network.

Paint, brushes and walls

The day referred to at the beginning of this chapter, which was the day on which Andrea began to paint, she went to an art store, bought orange, yellow, white, red, blue, black, light and dark green acrylic art-paints, one thick brush and about five other brushes of various sizes and forms from flat to round. Andrea did not buy canvases, however. She drove home, turned on some music and began to paint on the walls of her house.

Having bought the mediational means of paints and brushes in an art store distinguishes the action of painting on a wall from, for example, having bought acrylic wall paint and paint brushes at a home improvement store. Clearly, Andrea meant to produce art, not to paint her walls. Acrylic art paint and art brushes also *behave* differently on walls from acrylic wall paint and paint brushes. Thus, the choice of mediational means, the kind of paint and brushes Andrea decided to use, helped to produce the painting on her walls as art work.

The walls, due to their sturdy and immovable materiality, lent extended permanency to Andrea's first painting. This painting was not something she could hide from visitors. It was positioned for everyone to see and portrayed her as bold in her endeavour of producing art.

The mediational means, the paint and brushes, were not only used by Andrea to produce the art, but they were placed in her house, visible to everyone who came to visit. Thus, the paint and brushes, as well as Andrea's *Scream*, became frozen actions. All of these mediational means, easily interpreted as interconnected, helped produce Andrea as an artist.

Andrea's *Scream*

> An art which ever increasingly contains reference to its own history demands to be perceived historically; it asks to be referred not to an external referent, the represented or designated 'reality', but to the universe of past and present works of art.
>
> (Bordieu, 1984: 3)

Andrea painted her own version of the well-known painting *The Scream* (*Der Schrei der Natur*) by the Norwegian artist Edvard Munch. With her painting, Andrea referred back to late nineteenth- and early twentieth-century art. Edvard Munch had painted a variety of paintings depicting *The Scream* in various ways; and when Andrea paints her own version, as depicted in Figure 9.1, she displays her knowledge of these paintings, historically referring to past works of art. Because *The Scream* is so well known, painting a variation on the theme links Andrea to Edvard Munch; and Andrea's *Scream* in frozen form as viewed by the social actors in her network link Andrea's painting to art.

Yet, not all paintings that establish such a direct link to a prior work of art are necessarily viewed by other social actors as creative art work. Often – and I have examples from my ethnographic studies which I am not able to discuss in this chapter because of space limitations – paintings that establish a direct link to a past work of art are simply viewed as copies and are often seen as *not-really-creative* creative (frozen) action. In those cases, the copies are seen as imitations.

This leads us to the question *why was it different in Andrea's case?* and we find the answer in the fact that Andrea's *Scream* did not only directly link to Munch's paintings, but also directly linked to Andrea's own life. The painting was a depiction of how Andrea felt at the time: the painting displayed her own feelings of fear and anxiety due to the impending divorce. As Bilstein (2009: 15) notes, 'Kunst [ist] immer wieder Grundmuster für das Verstehen und Bewältigen der Alltagswelt . . . – vor allem dann, wenn es darum ging, Widersprüchlichkeiten und Unfassbarkeiten irgendwie

FIGURE 9.1 Andrea's *Scream*

zur Sprache zu bringen.' (Art [has] always [been] a system for understanding and realising the everyday . . . – especially then, when it is about explaining contradictions or the unfathomable.)

When social actors from Andrea's network came to visit, they not only saw the link to prior works of art, but also saw the relevance of the painting in Andrea's life. She had expressed her fear through her painting, communicating it more strongly than she had ever expressed her fear through words; and social actors who saw the painting understood, suddenly experiencing her fear as they viewed the painting. With this painting, Andrea had touched the viewers emotionally.

On the one hand, it may be this ability to communicate through the creative action that makes it *creative* in the sense of other social actors and Helga in Transcript 1; and it may be this ability to communicate through the creative action that makes the work produced a work of art. However, on the other hand, not every social actor who has the ability to communicate through paintings is necessarily viewed as an artist by others and also not necessarily views themselves as an artist.

The idea whether someone views themselves or is viewed by other social actors as an artist leads us to consider identity production. The question of identity here is: who produces Andrea's artist identity? Does Andrea agentively produce this identity herself, or is the identity prescribed to Andrea by others in her network? In other words, how does an artist become a community artist in everyday life?

Identity: Andrea, the community artist

Andrea's community artist identity was in part produced agentively through her use of the mediational means described above. In concurrence, the mediational means were utilised by members in her network. For example, Andrea utilised the narrative-skeleton (Transcript 9.2) to present herself as a person who has always wanted, but was not allowed to become an artist. The members in Andrea's network take on this narrative-skeleton and use it or refer to it, as Anna does in Transcript 9.4. While the narrative-skeleton had been initially introduced to other social actors by Andrea, and while the introduction, the telling and the retellings are agentively produced by Andrea, the narrative-skeleton also takes on form and function in the discourse used by others in Andrea's network. Through the telling, retellings, and references made by other social actors of Andrea's narrative-skeleton, the narrative-skeleton builds a part of Andrea's identity. The narrative-skeleton is thus sometimes used by social actors in Andrea's network to display their knowledge of and closeness to Andrea, especially if a newer member is present.

Andrea's community artist identity was also in part produced agentively through her acquisition and use of art books. The books and paintings within are often referred to by Andrea and are also mentioned in the discourse of others in Andrea's network. If someone in Andrea's network (or their children) has a question about art, they contact Andrea. She is the one in her network recognised for her knowledge.

Andrea's use of paint, brushes and walls, and finally Andrea's painting the *Scream* in part produced Andrea as an artist, but, here, the production of the artist was not agentive in the same sense as above. Andrea did use agency to buy the paint and the brushes. She did use agency to paint her *Scream* on a wall in her house. But she did not agentively produce her own artist identity. When we take another look at Transcript 9.3, we see that Helga views the painting as a work of art, while Andrea is not really sure. In fact, Andrea did not think about whether or not she was producing art at the time of painting the *Scream*, nor did she think about it afterwards. It was only when members of her network stated that it was art that Andrea made the link to her own narrative-skeleton as we saw in Transcript 9.3 as well.

But even here, where Andrea says in line (10) *hab ich ja immer gewusst,* (I've always known this,) and laughs, Helga reacts to this saying in lines (12) and (13) *nee,* (no,) *aber ehrlich!* (but really!). Even though her friend tells her that the painting is a piece of art, Andrea is unsure of its artistic value, which comes out in the (slightly ironic) tone of her utterance which in turn is reflected in the reassurance by Helga.

Andrea had painted the *Scream* in October 1999 because she felt a need to express her own feelings somehow. She had felt an urge to paint, but was not agentively producing an artist identity at that moment. However, Andrea's artist identity, even though non-reflected and not consciously produced at the time, already resided in her historical body (Nishida, 1956). Andrea, with her interest in art, had internalised how to produce it, but was not consciously perceiving herself as an artist. When asked by the researcher at a later date, Andrea explains in Transcript 9.5:

Transcript 9.5: Andrea about perceiving herself as an artist

(1)	Researcher:	*als du den Schrei gemalt hast,* when you painted the Scream,
(2)		*hast du dich dabei als Künstler gefühlt* did you feel yourself to be an artist
(3)		*oder wie hast du dich gefühlt?* or how did you feel?
(4)	Andrea:	*nein* no
(5)		*ich habe mich zu diesem Zeitpunkt* at that time I did
(6)		*nicht als Künstlerin gesehen.* not see myself as an artist.
(7)		*ich habe einfach gemalt . . .* I simply painted . . .
(8)		*ohne darüber nach zudenken* without thinking about
(9)		*wer ich bin.* who I am.

As Transcript 9.5 illustrates, Andrea did not perceive herself as an artist when she painted the *Scream*. In fact, even when she painted many other paintings following the *Scream*, she did not perceive herself as an artist. Andrea simply painted because she felt the need to express herself. Her creative actions were product-focused, product-creating, a form of self-expression and produced for herself. Once she had finished the *Scream* which expressed the fear and anxiety she felt due to the divorce, Andrea did more paintings on the walls of her house as illustrated in Figure 9.2.

After the *Scream* she painted two figures on a wall behind a stove holding hands and going up in flames, and on the stove stood framed pictures of her two young boys. This time, there was no direct or indirect link to prior works of art as there had been with her *Scream*, but there was now a direct link to her own *Scream* through the mode of colour (colour images can be found at www.sigridnorris.com/publications.html).

The strong visual link in colour (predominantly hues of blue and red) and the placement on a wall also linked another painting of two happy children holding flowers (where the 'happy' is indicated primarily through the orange background). The figures are again painted in blue and this time the flowers have red petals. Again, we find a strong link to Andrea's life and the impending divorce, and, again, social actors in Andrea's network made the link to art, at the same time as they became strongly aware of Andrea's life on an emotive level. Thus, social actors in Andrea's network utilised both modes of thought described by Bruner:

> There are two modes of cognitive functioning . . . each providing distinctive ways of ordering experience, of constructing reality. The two (though complimentary) are irreducible to one another . . . a story . . . is judged for its goodness as a story by criteria that are of a different kind from those used to judge a logical argument as adequate or correct.
>
> (Bruner, 1986: 11–12)

While Bruner actually discusses types of literary analysis, his notions, I believe, are also helpful when investigating the production of art. While Andrea was not consciously producing herself as an artist, she in fact consciously ordered her experience and constructed her reality through her paintings. In turn, social actors in her network judged the paintings for their *goodness* and they judged the *logical argument* of Andrea to be an artist

FIGURE 9.2 Paintings on Andrea's walls

inherent in the mediational means of narrative-skeleton, art books, and painting, as correct. As Andrea's development as an artist was discussed within her network and social actors came to see the works of art, these social actors demonstrated that they valued her paintings.

Transcript 9.6: You can't take them with you

(1)	Petra:	*weisst du* you know
(2)		*die sind ganz toll* they are really great
(3)		*aber was machst du* but what are you going to do
(4)		*wenn du ausziehst?* when you move?
(5)		*die kannst du doch nicht mitnehmen!* you can't (emphasis) take them with you!
(6)	Andrea:	*weiss nich* don't know
(7)		*nicht mein Problem* not my problem
(8)		*egal.* doesn't matter.
(9)		*wer das Haus kauft* whoever buys the house
(10)		*kann ja drüber streichen* can paint over them

Interesting here is that Petra and many others in Andrea's network saw Andrea's paintings as works of art, yet Andrea did not. She painted because she felt the need to paint. Andrea was not worried about keeping the paintings. Andrea painted for herself to work through her emotional turmoil, and painting afforded her the opportunity to make sense of her life at the time. She agentively tried to come to an understanding of what was happening in her life and expressed it in her paintings. When later asked by the researcher when it was that she saw herself as an artist for the first time, she explained in Transcript 9.7:

Transcript 9.7: At my first exhibition

(1)	Researcher:	*wann hast du dich zum ersten Mal* when was the first time that you
(2)		*als Künstlerin gesehen* saw yourself as an artist
(3)		*oder gefühlt?* or felt like one?
(4)	Andrea:	*bei meiner ersten Ausstellung* at my first exhibition
(5)		*in Fühlingen,* in Fühlingen,
(6)		*bei der ich vier Bilder verkauft habe.* when I sold four paintings.
(7)	Researcher:	*wie lange war das* how long was that
(8)		*nachdem du den Schrei gemalt hast?* after you had painted the Scream?
(9)	Andrea:	*im März 2000* in March 2000
(10)		*war die Ausstellung* was the exhibition
(11)		*und im Jahr 1999* and in the year 1999
(12)		*im Oktober* in October
(13)		*habe ich den Schrei gemalt.* I painted the Scream.

While Andrea had been producing the narrative-skeleton for many years, while she had learned much about art, and while she agentively bought acrylic art paints, art brushes, and produced paintings to express herself, she did not agentively produce her own artist identity at the time. While Andrea was painting and not worried about a possible identity as an artist, members of her network saw her as a community artist from the very beginning when she painted her *Scream*.

However, while members of her immediate network repeatedly told Andrea that her paintings were pieces of art, and while they consistently told her that she really was an artist, Andrea did not take on the artist identity until members of a wider community acknowledged her as an artist during her first exhibition, which she refers to in Transcript 9.7. This wider community was primarily made up of individuals who were directly linked to members of Andrea's immediate network. Only when she was celebrated as an artist in this somewhat wider network, and when she realised that social actors whom *she* did not know valued her paintings, did she take on the artist identity. The artist identity, thus, was quite strongly prescribed to her by her network and the somewhat wider network which saw the *goodness* in her paintings. This brings us back to the question of a *creative* action.

Recognition of creative actions in a community: Recognition of an artist

Many social actors paint in their everyday lives, but only few of them are considered community artists. How is it, I have been asking in this chapter, that some actions are actions that produce *one* social actor as a community artist while the same kind of actions may not produce *another* social actor as a community artist. When we look at the analysis above, we can see that there were three more or less consecutive but also to some extent overlapping stages of action in a long-term process that led to the production of Andrea as a community artist.

Some of Andrea's actions revealed her as a community artist: the action of telling a narrative-skeleton in many various ways to many different social actors; the action of collecting, reading, learning, and explaining art books; the action of buying acrylic art paint and art brushes; the action of painting.

Some actions of her immediate network identified Andrea as a community artist: the action of telling, re-telling, and referring to Andrea's narrative-skeleton; the action of asking Andrea questions about art; the action of visiting Andrea to view the paintings; the action of evaluating the paintings as art; the action of talking about Andrea as an artist.

Some of the wider network's actions confirmed Andrea as a community artist: the action of social actors, whom she did not know, looking at her paintings; the action of evaluating the paintings as art; the action of buying some of her paintings; the action of talking about Andrea as an artist; and the action of mentioning Andrea's art in a community newspaper.

Only once the third stage had occurred, did Andrea take on the artist identity for herself. Thus, what we find here is that Andrea set out to present herself in a certain way that alluded to a possible artist identity through a narrative-skeleton and other mediational means for many years before she began to paint. This possible presentation was taken up by the immediate network that she belonged to, and the network began to stabilise and co-form the possible artist identity. Then, once Andrea had produced a product that fit into the possible artist identity production, members in her network immediately read the painting as a work of art. However, it is only at the point when social actors outside Andrea's immediate network viewed her paintings and evaluated them as works of art that Andrea took on the artist identity for herself.

Conclusion

In this chapter I have asked the question: *How is it that some actions are actions that produce one social actor as a community artist while the same kind of actions may not produce another social actor as a community artist?* The answer, as I have illustrated throughout, does not lie in a creative action alone, but lies in many mediated actions and their embedded mediational means as well as in the social actor's historical actions and positioning: The answer lies in the long-duration process. This process entails actions that oscillate between individual and groups. It is neither the individual that produces the self as the artist, nor, however, is it the immediate or the wider network. However, without the wider network producing an identity, the identity is not fully developed in the individual.

Just as Andrea only took on the artist identity for herself once the wider network had acknowledged her as an artist, this acknowledgement, I also argue, would not have come about without the other two stages of actions within this long-duration process occurring in some way. An identity, as the community artist identity in everyday life, is thus not simply produced by creative actions as such, but is produced through many different actions by the social actor, by members of the immediate network, and by members of the wider network.

However, rather than viewing these three stages as separate units, when performing a holistic analysis we recognise that these stages are all inter-linked and cannot be completely separated in everyday life. While we can find some to be consecutively ordered, others are overlapping, and Andrea's taking on the identity of a community artist is in fact circular: it had to be made possible by Andrea; was prescribed by the immediate network; but was only internalised by Andrea once a wider network had acknowledged her paintings as works of art. Thus, in an example such as this, only once all three stages come together can a *creative* action produce a social actor as a community artist.

References

Bilstein, J. (2009) 'Die Kunst der Lehre und die Lehre der Kunst' in *Koblenzer Hochschulschriften*, Liebert, W. A. and Metten, T. (eds). Universität Koblenz-Landau: Impressum.

Bourdieu, P. (1984) *Distinction: A social critique of the judgement of taste.* Cambridge, MA: Harvard University Press.

Bruner, J. (1986) *Actual minds, possible worlds*. Cambridge, MA: Harvard University Press.

Lemke, J. L. (2000) 'Across the scales of time: Artifacts, activities, and meanings in ecosocial systems'. *Mind, Culture and Activity*. 7(4) 273–290.

Norris, S. (2004) *Analyzing multimodal interaction: A methodological framework*. London: Routledge.

Norris, S. and Jones R. H. (eds) (2005) *Discourse in action: Introducing mediated discourse analysis*. London: Routledge.

Norris, S. (2011) *Identity in (inter)action: Introducing multimodal interaction analysis*. Berlin, New York: de Gruyter.

Scollon, R. (1998) *Mediated discourse as social interaction*. London: Longman.

Scollon, R. (2001) *Mediated Discourse: The nexus of practice*. London: Routledge.

Wertsch, J. V. (1998) *Voices of the mind: A sociocultural approach to mediated action*. Cambridge, MA: Harvard University Press.

PART FOUR

Discourse, creativity and technology

No contemporary treatment of the relationship between discourse and creativity would be complete without a consideration of the impact of digital technologies on creative practices and creative products. Not only have technological developments in the last two decades made available to a wide segment of the population means of production and distribution hitherto available only to a very few, dramatically altering the ways creative individuals and groups – from musicians to novelists to graphic artists – do their work, but they have also dramatically altered the way we understand discourse itself, calling into question traditional distinctions between, for example, written and spoken language, introducing new genres and new discourse practices, and challenging previously held notions about what constitutes a text and what constitutes participation in social interaction.

In considering the role of technology in creative practices and processes, however, it is important to heed the advice offered by Toolan in the first chapter of this book, that we avoid engaging in a kind of technological fetishism that assumes that practices like writing fiction using mobile telephones or remixing digital video are automatically creative simply because they involve the novel use of new technologies. To be considered creative, such practices must result in products or solutions to problems in which some measure of value – whether it be aesthetic, psychological, or social – can be discerned. It is also important to bear in mind van Leeuwen's concerns voiced in the previous section that, while new technologies open up possibilities for creative expression, they also sometimes limit them through things such as design constraints built into software programs.

At the same time, partially because tools such as computers and the internet force us to think of language and communication in new ways and, as these technologies develop, to adapt to a rapidly changing discursive landscape, and partly because these tools also foster the ability for people to disseminate their creative works so widely and comment upon or interact with the creative efforts of others so easily, the rise of new media has resulted in increased attention to and debate around many of the issues we have been focusing on in this book: issues of aesthetic value, of originality and intertextuality, of everyday creativity, of the cognitive and social processes around the production and interpretations of texts, of collaboration, and of the identity of the creative artist.

In might be argued that in the world of the 'read-write web' where ordinary people are compelled to constantly produce texts (in all manner of modalities) and to publish these texts on their blogs and on various social media sites, where the barriers to creative expression have become relatively low, where many find themselves part of active social networks which offer constant support and feedback for creative production, and where people must constantly compete for attention within an ever-gathering storm of discourse, creativity has, like never before, become an important form of social capital. While much of this everyday creativity involves the appropriation, transformation and 're-mixing' of other texts, this does not necessarily make it less creative, but may, to some extent, induce us to consider more seriously the degree to which all creative expression builds upon what has come before it.

This section presents four different perspectives on creative practices and products associated with digital technologies, each from a rather different theoretical perspective. The chapter by Thurlow which begins the section gives an overview of the creative linguistic practices of young people using computers and mobile phones along with an analysis of how the mainstream media construct these practices as deviant and destructive on the one hand, and, on the other, co-opts them in order to market products to members of this generation. Drawing on many of the concepts introduced in earlier chapters of this book, Thurlow makes the case that discursive creativity in the new media is 'often *poetic*, usually *playful* and always *pragmatic*'. His approach, however, significantly broadens the circumference of the debate on what constitutes *value* in creativity, introducing a more sociological and critical perspective. Creative value, he points out, is just as much about power and politics as it as about poetry, playfulness and pragmatism, and while the creative efforts of young people on the internet may gain them considerable cultural capital among their peers, their discursive capital is almost always devalued in the symbolic marketplace controlled by adults. In response to this he offers a challenge to the discourse of creative value itself, suggesting that the role of scholarly examinations of creativity is not just to pass judgement on the value of particular creative products or creative practices but also to 'uphold and protect creative practice in/from a symbolic marketplace committed to the formulae of good-or-bad and right-or-wrong'.

Gillen's chapter in many ways complements and supports Thurlow's argument through an illustration of the creative practices of young people and adults learning about archaeology in an online virtual world. Drawing on theoretical paradigms developed in the New Literacy Studies, she sees creativity as an interactive process through which people work together to select, combine and adapt the various resources available to them to cope with the affordances and constraints of new situations, new environments and new tools for communication. In this respect, creativity is always 'emergent' as communicators negotiate the use of new communication tools or the adoption of new communication practices. Perhaps Gillen's most important contribution, however, is her insight that exploring new kinds of creativity requires the development of new kinds of creative research methodologies which can capture the range and diversity of practices, activities and experiences involved in technologically mediated creativity distributed across time and space. The approach that she develops, which she calls *virtual literacy ethnography*, results in an analysis which combines narrative and personal reflection with empirical observations based on the analysis of texts.

The chapter by Morrison and his colleagues takes a very different approach to the analysis of the relationship between technology and creativity. Rather than focusing on the creative practices people engage in when they use technology, they focus on the creative discursive processes which go into the development of technology itself. Through a reflection on the iterative, cooperative and participatory process of designing and trialling a mobile application for collaborative fiction, they examine not just how communication tools can be better designed to facilitate collaborative creativity, but also how the design of such tools is itself a collaborative creative process. They call their framework for the design of communication tools 'discursive design', an approach that sees practices of shared meaning-making at the heart of what are too often seen as strictly engineering problems. Studies of discourse and technology, they point out, often ignore the discursive processes that go into the design of technology, and in this chapter they make a persuasive case for further research in this area.

The final chapter in this section and in the volume is an analysis by Jones of the creative use of digital video by urban skateboarders in which he combines mediated discourse analysis with insights from complexity theory. His focus is less on the form of the creative products these young people produce or on the processes they go through when they are producing them and more on the kinds of social actions and social identities these products and processes make possible. Acts of creation, he argues, are ways in which human beings productively coordinate the complex multiple processes they undertake on different timescales – they are the way we turn the momentary actions of our day-to-day lives into meaningful life narratives and coherent identities. In this regard, creativity is seen as having a central role in individual and group learning, a point also pursued by Gillen in her chapter.

Although all of the chapters in this section, then, deal with ways in which creativity is mediated through technological tools, none of them focuses solely on technology, but they rather use the phenomenon of technological mediation as a starting point to raise larger issues about discourse and creativity. Creativity mediated through new technologies is in many ways not particularly different from that mediated through old technologies in terms of processes of combining existing resources to create new meanings. For a variety of reasons touched upon above, however, new communication technologies invite us to take a new perspective on these age-old processes.

CHAPTER 10

Determined creativity: Language play in new media discourse

Crispin Thurlow

> Creativity carries with it associations of freedom, agency and the unshackling of constraints. This is a beguiling but misleading view because creativity does not operate, unbounded, in an autonomous fashion. It is usually shaped by convention. It is about giving form to the material we draw on and transform, and this cannot be done without reference to existing rules, devices, codes and procedures.
>
> (Negus and Pickering, 2004: 68)

> Every culture proliferates along its margins. Irruptions take place that are called 'creations' in relation to stagnancies. Bubbling out of swamps and bogs, a thousand flashes at once scintillate and are extinguished all over the surface of a society. . . . Daily life is scattered with marvels, a froth on the long rhythms of language and history that is as dazzling as that of writers and artists.
>
> (de Certeau, 1997: 139–42)

In this chapter, creativity is examined in a less lofty sense than usual; it is not concerned with the elite, 'high culture' creativity of art and theatre or the commercial creativity of pop music and advertising. I want instead to consider a mode of creative practice that is no less artful, dramatic, lyrical or ubiquitous, but which is realised in the banal spaces and everyday exchanges of ordinary language users (cf. Carter, 2004; Cook, 2000; Crystal,

1998). I am also interested in the creative practices of a particular 'community' of *non-elites* – young people – who almost never control the official mechanisms by which they come to be represented, and whose discursive capital is almost always devalued in the symbolic marketplace (Cameron, 2000; cf. Bourdieu, 1991). In particular, I present a fairly straightforward demonstration of the kinds of creative linguistic, visual and interactional practices common to (but not necessarily typical of) a range of everyday new media. As a mode of vernacular literacy, discursive creativity in the new media is often *poetic*, usually *playful* and always *pragmatic*. My goal in reconvening these well-established principles from new media studies is to make two sets of observations: first, about the nature of creativity, and second, about the cultural politics of creativity in new media – particularly the playful use of language and technology by young people. I will start right away by sketching these 'observations' a little further and mapping them on to a broader literature.

First, new media discourse reveals perfectly one of the essential qualities (or realities) of creativity which Keith Negus and Michael Pickering note in the quotation above: its paradoxical dependence on normativity and structure. Creative practice always emerges out of the dialectical tension between fixity and mobility, constraint and freedom, convention and innovation, stricture and defiance, orthodoxy and heterodoxy, and, in the case of language, between 'grammar' and 'poetry' (cf. Jakobson, 1960; Maybin and Swann, 2006, 2007). There is no creative without normative. So much creative practice arises in the context of leisure, pleasure and play – in 'down time' rather than in 'work time' (Lieberman, 1977; Murdock and Ganim, 1993; Sutton-Smith, 2001). This is key for recognising how all creativity is marked as much by its *immediate* cultural and social context as it is by its messing with structures, systems, rules and codes. In discursive terms, creativity is as much a product of pragmatic *function* as it is of linguistic *form*. In the case of language play, creative practice is more often than not shot through with humour; this is language with serious interactional consequence (Bauman, 1984; Sherzer, 2002) but it is language often not intended to be taken seriously. Linguistic creativity (in the form of speech play or verbal art, for example) is also simultaneously extraordinary and everyday. These are creativities marked by their in-the-moment, *kairotic* quality rather than by any durable standards of measurability and orderliness (Harold, 2007). That a lot of creative practice is improvisational, non-instrumental and non-economic is precisely what can make it inconsistent with – and inconvenient to – the entrepreneurial priorities of neoliberal societies (Edensor *et al.*, 2010).

This brings me to the second observation I want to make: the political or ideological conditions under which *new media* creativity typically emerges. This is especially true for young people whose artful, pleasurable uses of new media language are frequently met with the censorious interventions of *language workers* (Thurlow, 2007) such as journalists and the commercial dealings of *semiotic merchants* such as advertisers (cf. Giroux, 2000; Jones and Schieffelin, 2009; Miles, 2000; Thurlow, 2006). For the most part, young people's creativity is misrecognised – if it is noticed at all – and they are denied an uncommodified, uncensored space for play. Even today, adult commentators sometimes appear reluctant to acknowledge or take seriously the inherent creativity and emerging literacies of new media discourse (Buckingham, 2007; Burgess, 2006, 2010; Kress, 2003). This reaction does not, of course, necessarily stop young people. On the contrary, by no means a homogenous 'community', many young people continue to create, to play, to communicate in spite of and perhaps *because* of the reactionary stance taken by adults. That some people's creative discourse comes to be marked as 'good' or 'acceptable' while others' is dismissed as 'unacceptable' or 'bad' is a simple fact of sociolinguistic life (Cameron, 1995; Foucault, 1981). This does not, however, render concern/critique unwarranted or pointless. Besides, as Michel de Certeau (quoted above) makes clear, it is precisely the unsanctioned, non-elite, unorthodox creations of everyday life – its 'flashes' and 'marvels' – that keep language and culture from stagnating. It is in this way too that, for Norman Fairclough (1992: 97), 'orders of discourse' come to be transformed.

Disciplining play: Language workers and 'anti-creative' censorship

> Teen Texting Soars: Will Social Skills Suffer?
>
> (*National Public Radio*, April 2010)

This news headline and its accompanying image in Figure 10.1 represent yet another moment in the seemingly endless public – and intensely mediatised – 'debate' about the deleterious impact of new media on young people's language and communication. (A 'debate' in scare quotes since one major party is seldom invited to speak.) In this case, some new but largely unremarkable findings from a Pew Foundation research report (see Pew, 2010) were given a quite unwarranted and negative spin since nothing in the

FIGURE 10.1 Teen texting soars
(photo © istockphoto)

original research report suggested social skills were 'suffering'[1]. To complement this lopsided framing of the facts, the news report was also visually framed by its decontextualised *conceptual image*, staged in such a way as to prioritise the technology, to disembed it from its cultural practice, to dismember its users and to exaggerate a sense of encirclement or closed-offness. This image-bank-like picture would not look out of place in an advertisement for any new technology – like the iPad – where marketers are understandably keen to highlight the shiny surfaces of the instrument rather than the grubby particularities of its uses and users.

Since publishing the results of a study into this type of biased reporting and its metalinguistic frenzy (see Thurlow, 2006, 2007), I regularly ask my students to track down more up-to-date examples of similar coverage. (In fact, the NPR report above was brought to my attention by my student Joy Lehnis.) I do this partly because people often suggest to me that the public is surely less anxious nowadays about the impacts of texting, instant messaging and social networking; 'Surely', they suggest, 'we know it's not the end of either language or the world as we know it?' 'Surely, people's attitudes are changing towards new technologies?' Evidently not. Here is a random sample of some surprisingly recent English-language news headlines:

- The death of English (LOL)
- IM worms its way into schoolwork

- Is our language changing for the worst?
- It's no laughing matter as cyber-slang rewrites teen literacy
- Secret of teen talk: The bizarre language of our texting generation
- Cellular alert: As texts fly, danger lurks
- Parents fear online chatting ruins kids' language skills
- Technology leaves teens speechless

The idea that young people are in any way 'speechless' is, of course, patently absurd. Young people, it seems, are fair game. The standards by which their new media practices are judged are certainly not those used to assess moments when elites and adults take up the technologies (see Thurlow and Bell, 2009). Apparently, when grown-ups break with convention and play with new technologies, it is something to be celebrated. When young people do the same thing, however, their creative output is reduced to *hieroglyphics*, *gobbledegook*, *technobabble*, *cryptic chat*, *jumble*, *ramblings*, *cryptic symbols*, *gibberish*, *argle-bargle* and *hodgepodge communication* to quote some of the media's terms of preference (see Thurlow, 2006).

Even as yesterday's young people are becoming today's young adults, and as 'new' media gradually become old media, this oppositional or polarising discourse appears quite tenacious. A number of important facts about young people's new media language are frequently left out of the picture, literally and figuratively; for example: the wider cultural practices that emerge from technology use; the complex embedding of new technologies into people's everyday lives; the variability of usage patterns and of linguistic style choices; and the relative ordinariness of the communication being done. What also gets overlooked or deliberately ignored is the playfulness and creativity of so many of the mediated exchanges. For already over-invested language workers like journalists, the leisurely, pleasurable rhythms of everyday life do not, of course, make for good or easy news. The news cycle accommodates the flash of revolution far more easily than it does the tedium of evolution.

Monetising play: Semiotic merchants and the commodification of creativity

It is not just language workers who exert their disciplinary gaze on new media language. Perhaps even more invested in the facts (and fictions) of it all are the quintessential semiotic merchants of commerce: advertisers.

It is here where the 'politics of culture' in young people's communicative practices is at its fiercest (Giroux, 2000). Within this cultural terrain, language is a major site of struggle. Indeed, critical discourse analysts have for a time been tracking the ways that human communication is becoming increasingly 'technologised' and commodified – valued, for example, as a matter of employability before citizenship and domestic life (e.g. Fairclough, 1992; Cameron, 2000). Primarily to satisfy the particular demands of the workplace and of 'enterprise culture', otherwise complex interpersonal processes are rendered by policy makers and educators as discrete, quantifiable *skills* or competences rather than as meaningful, situated, power-embedded *social* practices. And for a world of *work* there is not a lot of time given over to play. No wonder, then, that social networking, microblogging and text-messaging – all deployed primarily as technologies of sociability – are often devalued and sometimes derided. No wonder also that little value is attached to either play as a source of creativity (see Lieberman, 1977; Sutton-Smith, 2001) or to non-strategic creativity itself. There is little recognition in the kinds of technocratic policy statements above that much of what young people are actually doing with new media is already servicing interpersonal relations – 'cooperating with others', 'listening actively', 'resolving conflict', etc.

None of this, of course, means that young people's new media discourse is without certain value to the market. Although they have sometimes been caught unawares by the uptake and development of new technologies (text-messaging is a case in point), people of commerce are always quick to catch up and their drive to capitalise is voracious. (As I learn from an AdSense promotion in my *iNeedle* blog, it's all about 'monetisation'.) Excited claims made in the 1990s for the democratising, emancipatory and participatory potential of new media have had to be sharply revised as the involvement of commerce has intensified. Politicians, too, know – just as Barack Obama did – that new media are vehicles to voters.

In this specifically instrumental term, the playful and creative practices of new media language are exploited as resources by which adults not only demonise but also commodify youth (Giroux, 2000; also Miles, 2000). The idealisations and misrepresentations of new media by the old media are ultimately focused on the preparation of young people for consumption – as both agents and objects, as people 'sold to' and as people used to sell to others. The trendy young 'creative-type' used widely to embody the Apple Mac brand exemplifies the way youth is thus 'transformed into a market strategy and a fashion aesthetic used to expand the consumer-based needs of privileged adults' (Giroux, 2000: 19). Other prime examples of this commercial refashioning are AT&T's much talked about (in the USA at least) *My BFF Jill* text-messaging campaign with its exaggerated communication breakdown between parent and child (subtitles for 'textese') and its equally implausible texting granny (see Jones and Scheiffelin, 2009).[2] The same basic commodification of new media language also appeared in Calvin Klein's aestheticisation of text-message style in its 2007 Canadian advertising campaign for the *ckin2u* perfume (see Figure 10.2).

FIGURE 10.2 Calvin Klein advertisement
(Image courtesy of The Aduertising Archives)

So it is that language workers and semiotic merchants work together to exaggerate the 'separateness' of youth culture in order to distinguish young people as a newsworthy topic and profitable market. In this way, too, young people and their creative practices become consumable by disembedding them from their ordinary playful contexts of use and recontextualising them for the glossy purposes of marketing (Thurlow, 2007). Young people's communication is effectively being sold back to them. In the process, their authentic, vernacular creativities are resemioticised as exotic and outrageous, foolish and pointless, offensive and menacing.

Playing with language: The creative tactics of new media discourse

> Instead of viewing language and culture as systems where everything holds together nicely and neatly, I see them as open systems with squishes, fuzziness, leaks, *inventions*, *constructions*, negotiations, and imaginations, and as constantly emergent. Discourse is crucial to the language-and-culture intersection, the locus of the *actualization of potentials* provided by both language and culture as well as personal experience. In this intersection, *creativity, imagination, and play* are essential. . . . Again, speech play and verbal art are at the heart of this process, and they help us to understand not only the relationships among language, culture, and thought but also the *creative spirit* which constantly inspires new forms of expression and *aesthetic creation*.
>
> (Sherzer, 2002: 10, emphasis mine)

In her study of new media language creativity, Sarah North (2007: 539) discusses the close relationship between creativity, humour and play; she cites the shared features identified by Murdock and Ganim (1993): spontaneity, release, incongruity and unexpectedness. In one of the definitive texts dealing empirically with new media discourse, Brenda Danet (2001) too demonstrates its inherent playful nature *and* creative spirit, confirming Joel Sherzer's observation (quote above) that speech play/verbal art are (a) at the heart of all discursive life; (b) hinge on the combination of creativity, imagination and play; and (c) transform and sustain both language and culture. It is the transformative potential of even the most apparently banal acts of creativity that forces a rethink of the 'news' media's often scornful, dismissive approach to new media. In this next section – and the one after – I want to look more concretely at the realities of new media discourse and to review the kinds of playful, creative practices that give rise to so much mediatised panic and that become the fetishised objects of commerce. In other words, to see what all the fuss is about.

I start with the following extract from Misao Oki, a student of mine, as part of a transcription twice as long which she completed in 2008 as a class assignment. This four-minute snapshot of an exchange between two university students chatting on AOL's AIM service late one weekday night exemplifies the kinds of ordinary, playful, highly sociable moments enjoyed by many (young) people online.

Extract 10.1 – 'just playin'

P1 = Filipino-American male / P2 = Japanese-American female

starts 12:39:39 am

1. P1: is that who you went to the passion concert with?
2. P2: YES
3. P2: woooww
4. P2: how did u know!?!?!?!?
5. P1: lol you told me . . ?
6. P1: how did you hear about passion anyway?
7. P2: paul
8. P2: paul invited me. and i listend to passions song on youtube
9. P2: and i LOVED it
10. P1: oh haha
11. P1: haha is he flip
12. P2: i love acoustic guitar
13. P2: and piano
14. P1: i knowwwwww
15. P2: i have a picture with him hehe and his songs on my ipod
16. P1: no is paul flip?
17. P2: nope
18. P2: hes thai
19. P2: ya i felt so out of place at the concert
20. P2: hahaha
21. P1: lol yeah you stuck out like a sore thumb =P

22. P2: aww really!? sad
23. P1: just playin *[P2 name]*
24. P1: hahaha
25. P2: oh ok
26. P2: made me feel very sad for a sec
27. P1: haha jus for a sex
28. P1: sec
29. P1: freak..
30. P1: i always do that
31. P2: ahaha nnhmmm i bet you did that on purpose
32. P1: lol mmhmm i did

ends 12:43:39 am

As P2 puts it himself, the two of them are 'just playin' (line 23) here, hanging out together. This interactional frame and/or key is maintained throughout by the pace of turn-taking as well as by the following discursive tactics:

- *topics* (gossip, concert going, mutual friends)
- *humour* (line 15 'i have a picture with him hehe')
- *laughter* (line 5 'lol', line 10 'haha', line 11 'haha', line 15 'hehe', line 20 'hahaha', line 21 'lol', line 24 'hahaha', line 27 'haha', line 31 'ahaha', line 32 'lol')
- *taboo* (lines 26–28: punning on 'sex' for 'sec')
- *in-group terms* ('flip' for Filipino in lines 11 and 16)
- *teasing* (line 21: 'you stuck out like a sore thumb' – a reference to P2 Japanese ethnicity in a predominantly Filipino concert audience; line 29 'freak')
- *bald-on-record responses* (line 5 'you told me'; line 14 'i knowwwwww'; line 20 'hahaha'; line 31 'I bet you did that on purpose')

This 'space for play' is clearly also grounded through the informal register (cf. North, 2007) established primarily by the following non-standard/ innovative orthographic tactics:

- *'incorrect' spelling and grammar* (e.g. P2's use of an uncapitalised 'i'; also line 8 'passions' for Passion's' and 'listend' for listened; line 15 'songs' for song's; line 18 'hes' for he's)

- *colloquialisms* (line 17 'nope'; line 23 'playin' for playing; line 26 'sec' for second; line 27; 'jus' for just)

The performative exactment of this playful frame is also evidenced in various other stylistic choices made for exclamatory or affective impact:

- *onomatopoeia and capitalisation* (line 2 'YES'; line 3 'woooww'; line 9 'LOVED'; line 14 'knowwwwww'; line 22 ' aww'; line 31 'nnhmmm'; line 32 'mmhmm')
- *emotes* (line 21 '=P' for awkward tongue stuck out; line 22 'sad')
- *reduplicated punctuation* (line 4 '!?!?!?!?'; line 5 '..?'; line 22 '!?'; line 29 '..')

On the surface, this sample of instant-messaging epitomises the kind of non-standard 'shorthand' language which the media loves to hate. Certainly it proves that some new media discourse does indeed look like this; however, not all of new media language looks like this (there are even stylistic differences between P1 and P2), and individuals may not write the same way all the time (style is inevitably shaped by audience design; Bell, 1984/1997). Regardless, the point here is that the language is hardly 'hieroglyphic' – even to a complete outsider. There are moments which may well be missed (e.g. the band Passion, its largely Filipino following, or the =P emoticon), but even the density of unorthodox spelling and non-standard orthography does little to obstruct understanding. This is partly because, as Tim Shortis (2007) notes, new media language play – what he calls 'vernacular orthography' – often entails the recycling of existing forms and familiar practices. Furthermore, the real significance (and, indeed, creativity) of new media language does not lie simply in its orthographic innovations for *brevity and speed* or for *prosodic marking/speech simulation* (see Shortis, 2007; Thurlow and Poff, 2012). Other modes of discursive creativity are always at play, and new media discourse is never simply about form. There is certainly *type play* (e.g. capitalisation, repetition) and *word play* (e.g. punning), but there is also *interactional play* (e.g. teasing, game-playing and mock combat), *identity play* (e.g. nicks, gender-swapping), *sound play* (e.g. accent stylisations, letter-number homophones), and *topical play* (e.g. experimenting with norms of appropriacy and politeness). Creative practice is motivated by – and expressed in – the pragmatic, relational needs of participants as much as it is by their deliberate attempts to play with the norms of spelling, punctuation and grammar.

In these terms, new media discourse exemplifies perfectly the creative possibilities that emerge out of, in spite of and *because* of restrictions and obstacles. The following short extracts (10.2 to 10.7) offer another example of the broader discursive (as opposed to 'linguistic') creativities of new media.

These are drawn from Graham Jones and Bambi Schieffelin's (2008, 2011) work on the quotative *like* (e.g. 'And I was like . . .') used for *reported speech* in instant messaging.

Extracts 10.2 to 10.7 – The play's the thing

Extract 10.2

1.	sabrina461:	i asked him a question and he
2.		was like well . . . your bf is a dumbass
3.	sabrina461:	and i'm like oooo ok

Extract 10.3

1.	FuManChewy:	im like, zachary u okay?
2.	FuManChewy:	zachary?
3.	FuManChewy:	ZACHARY?
4.	FuManChewy:	omg wtf

Extract 10.4

1.	FuManChewy:	and i was like !!!!! bastard!

Extract 10.5

1.	Kitty59:	i hope matthew young sees it and is
2.		liek OOOOHHHH:-D

Extract 10.6

1.	FuManChewy:	and he's like::moans:: Nooo

Extract 10.7

1.	CaliSuzi:	hahaha wouldn't that be depressing if they were
2.		lie "we're taking you to times square!!!"
3.		*complete silence*

I borrow these snippets here less for the quotative *like* and more as examples of their dramatic staging (or re-staging) of playful exchanges. Of course, all the way through, there are the usual deployments of non-standard/vernacular orthography. What is more interesting is the use of this *formal* creativity for creating a sense of *functional* immediacy and dramatic tension. Note, for example, the dramatic pause '. . .' and the hyper-stylised exclamation 'oooo' in Extract 10.2 (lines 2 and 3), the frustrated call out 'ZACHARY?' in Extract 10.3 (line 3), the '!!!!!' and 'OOOOHHHH' for shocked or surprised phonological effect in Extracts 10.4 and 10.5 (line 2), the exaggerated performance of '::moans::

and 'Nooo' in Extract 10.6, and, finally, the dramatic staging of '*complete silence*' in Extract 10.7 (lines 2 and 3). New media discourse always involves a combination of reflexive, meta-textual awareness and of a deeply social awareness of audience design; this is how communicators imaginatively bring the words off the screen and bring language into life. In many ways, they must also resist and rework the deep-seated practices of a lifetime. This brings me to the next point I want to make about new media creativity.

Beyond language play: The textual pleasures of multimodal design

> Learning how to write is unlearning how to produce multimodal, visual meaning representations. This, it should be underscored, is an *ideological* process. Every written document is a visual document . . . Thus, visuality is not lost in practice, but it is lost in *ideological conception* of the writing and reading process.
>
> (Blommaert, 2005: 116)

Most self-respecting scholars interested in new media discourse recognise that it is no longer sufficient to fixate on the linguistic form or orthographic features of, say, texting, emailing and messaging (see Androutsopoulos, 2000; Jones, 2011). To do so means otherwise doing little to rise above the superficial, prurient curiosity of journalists and advertisers. It also means that the broader, richer creativities of new media discourse are similarly misrecognised and undervalued. As Jan Blommaert remarks in the quote above, judgements about language and creativity are invariably matters not of actual practice but of ideology, status and inequality.

New media discourse is seldom just a simulation of speech; it is also an expressive, creative mode in its own right, with its own meaning potentials, its own aesthetic and poetic pleasures. Nor is it only academics who recognise that 'semiotic modes other than language are treated as fully capable of serving for representation *and* communication' (Kress and van Leeuwen, 2001: 46). New media discourse hinges on the reworking of material/design constraints, the mixing of genres, the hybridising of speech/writing. Along the way, new *translocal* and multilingual geographies are also being imagined (Leppänen *et al.*, 2009; cf. also Androutsopoulos, 2000) and innovative ways of experimenting (and combining) with both the *modes* and the *media* available to communicators.

Textual 'mashups' like *Facebook* profiles demonstrate quite pointedly the convergence of new and old media and the intertextual formats of new media. They also highlight a general decentring or, at least, repositioning of language, none of which is to say that the written words no longer matter; rather, that the range of communicative resources competing for our attention is greatly expanded. As the technological capabilities are changing, so are the semiotic and creative possibilities. Not only does it become increasingly difficult to neatly locate language, but it is harder and harder to understand its relative significance for 'readers'.

Despite the excited rhetoric surrounding so-called Web 2.0 technologies such as social networking sites, 'interactivity' and multimodality are not new to new media. As Blommaert says, all writing (or typing) is unavoidably a *visual* document; as such, *all* new media – text-based or otherwise – have *always* been multimodal just as they have always been interactionally and culturally situated. Indeed, Tim Shortis (2007) lists some of the long-standing graphical tactics used in text-based new media: the use of colour, photos, sound; experiments with textual design/layout; code play with dingbats and webdings (e.g. □✕ⓘ?🚨ⓘ● for 'crispin'); 'eye dialect' (e.g. tuff for 'tough'); the kind of reduplication already seen above (e.g. !!!!????); and typographic rebuses (e.g. @}-'-,--- for a rose). Similarly,Brenda Danet (2001) offers a 'historical' account of ASCII art from the 1990s – pictures created manually using only the 95 printable key-strokes in the *standardised* ASCII system.

Already in Extracts 10.1 to 10.7 above a quite self-evident visual or multimodal literacy is at play. Along these lines, and lifted from the work of Marisol Del-Teso-Craviotto (2008), Extracts 10.8 and 10.9 prove yet again the deeply creative and fully communicative effects of these kinds of typographic invention in even the most restricted new media contexts.[3]

I highlight these two particular exchanges between participants in online chatrooms for their obvious creative visuality but also for their tangible sensuality. The sexual exchange is self-consciously playful but fully sexual in its illocutionary and, most likely, perlocutionary force. Although it is, in theory, no less true of all online discourse, these two extracts manifest perfectly the shift from performance to performativity, from the metaphoric/virtual/symbolic to the literal/real/physical. In other words, the '(@) (@)' in Extract 10.8 and the '//U\\' in Extract 10.9 playfully turn typeface into image also whilst having serious intent of turning people on. Not only are these practices fully multimodal (i.e. moving well beyond the basics/limits of language), they are also *material* in such a way that the language is *felt*, is *lived*, is *organic* (see Jones, this volume and 2011). In these terms, but also in more ideological terms, language and especially linguistic creativity can have physical consequences (C.F. Thurlow, 2011). It is this that makes new media discourse both pleasurable and political.

Extract 10.8 and 10.9 – Online sex play

Extract 10.8 (AOL chatroom, Lesbian 30s)

1.	HOTAZHEAT88:	COME HERE I GOT TITS (@) (@)
		. . .
2.	Xboigyrlx2227:	OOOOOOO NICE AND PERKY AND FIRM
3.	Da1nonlyteas:	my god look at the size of Hots nips
		. . .
4.	Xboigyrlx2227:	SEND THEM THIS WAY

Extract 10.9 (AOL chatroom, Gay 30s)

1.	KNIGHT OWL 928:	i wanna meet
2.	SameOldTrik:	i wanna 'MEAT' somebody too
		. . .
3.	SameOldTrik:	anybody naked right now . . .
4.	RHNIC 21:	i am
5.	reb3579reb:	me
6.	RujolikidR:	oh yes
7.	KNIGHT OWL 928:	me
8.	SameOldTrik:	me 2
		. . .
9.	SameOldTrik:	b naked & jerk it boys
		. . .
10.	SameOldTrik:	//U\\ play w/this *penis between legs*

For play: New media language and/as vernacular creativity

> The transfer of texts from one environment to another also causes shifts in judgments of creativity . . . and thus also in the capacity to be perceived as creative.
>
> (Blommaert, 2005: 106)

FIGURE 10.3 Language play: Concrete poem by Michael P. Garofato

> Creativity and innovation always builds on the past. The past always tries to control the creativity that builds on it.
>
> (Lessig, 2002: np)

As Jan Blommaert notes above, being misunderstood and misrepresented outside the spaces and moments of creation is an inevitability. Creativity is always constrained and *determined* in this way. By the same token, it is the refashioning of impediments, the challenging of norms – the reworking of the old, the conventional, the hegemonic – that makes things generative and transformative. This is not, however, to say that creative practice submits compliantly to 'the past'. An equally *determined* effort is needed to uphold and protect all manner of creativities in/from a symbolic marketplace committed to the formulae of good-or-bad and right-or-wrong. Creativity is determined, and creators must therefore be determined as well.

So few spaces nowadays remain completely beyond the scope and disciplinary power of advanced capital. So little creative freedom, so few uncommodified spaces for play (Lessig quotation above; also Burgess, 2010). The boundaries between art, commerce and vernacular creativity are increasingly blurred. Nowhere is this more apparent than the 'incorporation' of the work of graffiti artist Banksy; his unauthorised, 'on-the-run' moments of playfully reinventing fixed resources have been legitimated and canonised by, for example, a 2009 exhibition of his work in Bristol's City Museum and Art Gallery.[4] The vernacular becomes elite, the creative becomes normative. In these terms, the cultural politics of young people's new media creativity bears many similarities to that of graffiti. It is usually judged as unauthorised and subject to ridicule or outright condemnation (Thurlow, 2006, 2007). New media discourse is similarly playful, artful and driven by a provocative but also aesthetic regard for normative practices such as grammar, spelling and orthography, as well as interactional norms and other conventional standards of discursive 'appropriacy'.

A similar double standard is used to evaluate the creativity of new media language; either it is framed by exaggerated novelty and excitable 'presentism' or by alarmist opposition and the colonising literacies of established (i.e. adult) elites. Even within academic writing, it is hard to shift discussion beyond oversimplified dichotomies of authorised and unauthorised practice: for example, in framing certain discursive practices as *anti-language* (Halliday, 1975), *subcultural* (Androutsopoulos, 2000) or *transgressive* (Scollon and Scollon, 2003). Dwight Conquergood (1997: 354–5) offers another seemingly attractive, but no less problematic, notion of *counterliteracy* which 'challenges, mimics and carnivalises . . . textual power . . . and literate bureaucracy'. New media discourse would also seem suited to the title of 'weird English' (Ch'ien, 2004) since it expresses experimental, unorthodox and perhaps decolonising acts of identity. However, I worry that these otherwise respectful, even celebratory, rhetorical moves run the risk of reinscribing the already dichotomised, normative economies of both literacy and creativity (cf. Blommaert, 2005; Edensor *et al.*, 2010). Is there a way out of this?

It is certainly unfortunate that the range and diversity of 'new literacies' (Kress, 2003) evident in new media language is so often misrepresented – by teachers, parents, journalists and so on. Just as technologies are inevitably embedded in complex ways into people's everyday lives, however, new media creativity is always *integrated* practice and not something that sits outside some detached 'standard' practice. To judge new media language against the standards or norms of formal discourse – 'correct' grammar or 'proper' language – is both nonsensical and ideological (cf. Irvine and Gal, 2000; Blommaert, 2008). Rather like Banksy, the artist Jenny Holzer is fully sanctioned to play with language for the elite practices of 'high culture'. She is even permitted to project her verbal art across the walls and surfaces of cities around the world (cf. Jaworski, 2010).[5] Compare this with the creative practices of 'unincorporated' graffiti artists whose work is usually dismissed and prosecuted as 'meaningless' vandalism (Pennycook, 2010). The language play and verbal art of new media communicators is certainly no less creative, no less imaginative, no less reflexive than, say, concrete poetry or other verbal art. Nor is new media language any less significant for language and culture (Blommaert, 2005; de Certeau, 1997; Sherzer, 2002). Unlike other orthodox or authorised modes of creativity, the language of new media is an expression of *vernacular* creativity (Burgess, 2006, 2010; Edensor *et al.*, 2010). In contrast to the instrumental neoliberal uses of 'creativity' (as in the 'creatives' of marketing discourse and in other so called 'creative industries'), the vernacular creativities of new media are a form of what Paul Willis (1998) calls 'grounded aesthetics' characterised by their playful, pleasurable nature. These forms of symbolic creativity are neither extraordinary nor spectacular, but rather mundane and intensely social. And their transformative potential lies precisely in their being non-elite, non-institutionalised practices.

Notes

[1] The NPR story is available (22 September 2011) online at <http://www.npr.org/templates/ story/story.php?storyId=126117811>.

[2] It is not that the granny is texting that is implausible but her use of an exaggerated 'textese' style.

[3] The material from Del-Teso-Craviotto is not strictly from young people but there is no reason to think that young people are not engaged in similar exchanges. Besides, this material also debunks the popular misconception that new technologies are necessarily or somehow exclusively a 'teen thing'.

[4] See http://www.guardian.co.uk/artanddesign/gallery/2009/jun/12/banksy-bristol-art-exhibition

[5] For Jenny Holzer, see http://www.jennyholzer.com

References

Androutsopoulos, J. K. (2000) 'Non-standard spellings in media texts: The case of German fanzines'. *Journal of Sociolinguistics* **4** (4) 514–33.

Bauman, R. (1984) *Verbal art as performance*. Prospect Heights, IL: Waveland.

Bell, A. (1984/1997) 'Language style as audience design' in *Sociolinguistics: A reader and coursebook*. Coupland, N. and Jaworski, A. (eds). New York: St Martin's, 240–50.

Blommaert, J. (2005) *Discourse: A critical introduction*. Cambridge: Cambridge University Press.

Blommaert, J. (2008) *Grassroots literacies*. London: Routledge.

Bourdieu, P. (1991) *Language and symbolic power* (J. B. Thompson, trans). Cambridge, MA: Harvard University Press.

Buckingham, D. (2007) *Beyond technology: Children's learning in the age of digital culture*. Cambridge: Polity.

Burgess, J. (2006) 'Hearing ordinary voices: Cultural studies, Vernacular creativity and digital storytelling'. *Journal of Media and Cultural Studies* **20** (2) 201–14.

Burgess, J. (2010) 'Remediating vernacular creativity: Photography and cultural citizenship in the flickr photo-sharing network' in *Spaces of vernacular creativity: Rethinking the cultural economy*, Edensor, T. *et al.* (eds). London: Routledge, 116–26.

Cameron, D. (1995) *Verbal hygiene*. London: Routledge.

Cameron, D. (2000) *Good to talk? Living and working in a communication culture*. London: Sage.

Carter, R. (2004) *Language and creativity: The art of common talk*. London: Routledge.

Ch'ien, E. (2004) *Weird English*. Cambridge, MA: Harvard University.

Conquergood, D. (1997) 'Street Literacy' in *Handbook of research on teaching literacy through the communicative and visual arts*, Flood, J., Brice Heath, S. and Lapp, D. (eds). New York: Simon and Schuster, 354–75.

Cook, G. (2000) *Language play, language learning*. Oxford: Oxford University Press.

Crystal, D. (1998) *Language play*. London: Penguin.

Danet, B. (2001) *Cyberpl@y: Communicating online*. Oxford and New York: Berg.

de Certeau, M. (1997) *Culture in the plural* (trans. T. Conley). Minneapolis: University of Minnesota.

Del-Teso-Craviotto, M. (2008) 'Gender and sexual identity authentication in language use: The case of chat rooms'. *Discourse Studies* **10** (2) 251–70.

Edensor, T., Leslie, D., Millington, S. and Rantisi, N. M. (eds) (2010) *Spaces of vernacular creativity: Rethinking the cultural economy*. London: Routledge.

Fairclough, N. (1992) *Discourse and social change*. Cambridge: Polity.

Foucault, M. (1981) 'The order of discourse' in *Untying the text: A poststructuralist reader*, Young, R. (ed.). London: Routledge and Kegan Paul, 48–78.

Giroux, H. A. (2000) *Stealing innocence: Youth, corporate power, and the politics of culture*. New York: St Martin's.

Halliday, M. (1975) 'Anti-languages'. *American Anthropologist* **78** (3) 570–84.

Harold, C. (2007) *Our space: Resisting the corporate control of culture*. Minneapolis: University of Minnesota.

Irvine, J. T. and Gal, S. (2000) 'Language ideology and linguistic differentiation' in *Regimes of language*, Kroskrity, P. (ed.). Santa Fe, NM: School of American Research Press, 35–83.

Jakobson, R. (1960) 'Closing statement: Linguistics and poetics' in *Style in language*, Sebeok, T. A. (ed.). Cambridge, MA: MIT Press, 350–77.

Jaworski, A. (2010) *Language art: Performances of space and text*. Paper presented at the 3rd Linguistic Landscapes conference, Strasbourg, 5–7 May.

Jones, G. M. and Schieffelin, B. B. (2008) 'Enquoting voices, accomplishing talk: Uses of be + like in instant messaging'. *Language and Communication* **29** (1) 77–113.

Jones, G. M. and Schieffelin, B. B. (2009) 'Talking text and talking back: "My BFF Jill" from boob tube to YouTube'. *Journal of Computer Mediated Communication* **14** (4) 1050–79.

Jones, G. M. and Schieffelin, B. B. (2011) 'When friends who talk together stalk together: Online gossip as metacommunication' in *Digital discourse: Language in the new media*, Thurlow, C. and Mroczek, K. (eds). Oxford: Oxford University Press, 26–27.

Jones, R. (2011) 'C Me Sk8: Discourse, technology and 'bodies without organs' in *Digital discourse: Language in the new media*. Thurlow, C. and Mroczek, K. (eds). Oxford: Oxford University Press, 321–329.

Kress, G. (2003) *Literacy in the new media age*. London: Routledge.

Kress, G. and van Leeuwen, T. (2001) *Multimodal discourse: The modes and media of contemporary communication*. London: Arnold.

Lee, C. (2011) 'Texts and practices of micro-blogging: Status updates on Facebook' in *Digital discourse: Language in the new media*, Thurlow, C. and Mroczek, K. (eds). Oxford: Oxford University Press, 110–128.

Leppänen, S., Pitkänen-Huhta, A., Piirainen-Marsh, A., Nikula, T. and Peuronen, S. (2009) 'Young people's translocal new media uses: A multiperspective analysis of language choice and heteroglossia'. *Journal of Computer Mediated Communication* **14** (4) 1080–107.

Lessig, L. (2002) *Free culture*. Keynote presentation at the Open Source Convention, San Diego, USA, July 24, 2002. Available online (22 September 2011) at: <http://www.oreillynet.com/policy/2002/08/15/lessig.html>

Lieberman, J. N. (1977) *Playfulness*. New York: Academic Press.

Maybin, J. and Swann, J. (eds) (2006) *The art of English: Everyday creativity*. London: Palgrave Macmillan.

Maybin, J. and Swann, J. (2007) 'Everyday creativity in language: Textuality, contextuality, and critique'. *Applied Linguistics* **28** (4) 497–517.

Miles, S. (2000) 'Consuming youth' in *Youth lifestyles in a changing world*, Miles, S. (ed.). Buckingham: Open University.

Murdock, M. C. and Ganim, R. M. (1993) 'Creativity and humor: Integration and incongruity'. *Journal of Creative Behavior* **27**, 57–70.

Negus, K. and Pickering, M. (2004) *Creativity, communication and cultural value*. London: Sage.

North, S. (2007) '"The Voices, the Voices": Creativity in online conversation'. *Applied Linguistics* **28** (4) 538–55.

Pennycook, A. (2010) 'Spatial narrations: Graffscapes and city souls' in *Semiotic landscapes: Language, image, space*, Jaworski, A. and Thurlow, C. (eds). London: Continuum, 137–50.

Pew Research Center (2010) *Future of the Internet IV* http://www.pewinternet.org/Reports/2010/Future-of-the-Internet-IV.aspx

Scollon, R. and Scollon, S. B. K. (2003) *Discourses in place: Language in the material world*. London: Routledge.

Sherzer, J. (2002) *Speech play and verbal art*. Austin, TX: University of Texas.

Shortis, T. (2007) 'Gr8 Txtexpectations: The creativity of text spelling'. *English, Drama, Media* 8 (June 2007), available (05 April 2009) online at: <http://www.nate.org.uk/index.php?page=9&id=9>.

Sutton-Smith, B. (2001) *The ambiguity of play*. Cambridge, MA: Harvard University.

Thurlow, C. (2006) 'From statistical panic to moral panic: The metadiscursive construction and popular exaggeration of new media language in the print media'. *Journal of Computer Mediated Communication* **11** (3) 667–701.

Thurlow, C. (2007) 'Fabricating youth: New-media discourse and the technologization of young people' in *Language in the media: Representations, identities, ideologies*, Johnson, S. and Ensslin, E. (eds). London: Continuum, 213–33.

Thurlow, C. (2012) 'Speaking of difference: Language, inequality and interculturality' in *Handbook of critical intercultural communication*, Halualani, R. and Nakayama, T. (eds). Oxford: Blackwell, 227–247.

Thurlow, C. and Bell, K. (2009) 'Against technologization: Young people's new media discourse as creative cultural practice'. *Journal of Computer-Mediated Communication* **14** (4) 1038–49.

Thurlow, C. and Poff, M. (2011) 'Text-messaging' in *Handbook of the pragmatics of CMC*, Herring, S. C., Stein, D. and Virtanen, T. (eds). Berlin and New York: Mouton de Gruyter.

Willis, P. (1998) 'Notes on common culture: Towards a grounded aesthetics'. *European Journal of Cultural Studies* **1** (2) 163–76.

Acknowledgements

Ideas for this chapter were first presented at the *Roundtable on Discourse Analysis* in May 2009 at the City University of Hong Kong; I am very grateful for the all-round hospitality of the organisers and the collegial comments of my co-presenters. Thanks to Misao Oki for her permission to use the material in Extract 10.1 and to Michael P. Garofalo for permission to reproduce his concrete poem on page 184. All other images are used for scholarly commentary and/or criticism. Finally, special thanks are due to Rodney Jones and an external reviewer for their thoughtful and helpful suggestions.

CHAPTER 11

Archaeology in a virtual world: Schome Park

Julia Gillen

. . . The fire, and cracks
Of sulphurous roaring . . .
The Tempest I.ii

Creativity

In this chapter I will develop a sociocultural perspective on creativity through an examination of teenagers and adults studying archaeology in a virtual world as part of the Schome Park Programme (Gillen *et al.*, 2009; Twining, 2009).

Notions of creativity are profoundly cultural judgements, instantiated in discourses. In earlier ages, scholarly work shaped a discourse of creativity that stressed the artistry of the profoundly talented individual, exemplified by Samuel Johnson's *Lives of the Most Eminent English Poets* (1779–81). This discourse developed with the later Romantic tradition, and is still influential today. It imagines the creative person as a solitary individual, prototypically a man of genius working with great intensity on artistic products, perhaps paintings or poems. Their outputs are in some way innovative and thus deemed 'original', simultaneously recognisable in their cultural origins yet offering some reworking or evident new thinking. Valued products are each unique, preserved in acts of buying and selling and the lasting attention of scholars.

Considerable work in applied linguistics such as that by Carter (2004), Cook (2000) and Maybin and Swann (2006) has fruitfully extended perspectives on what may be considered creative linguistic acts. Linguistic creativity has been considerably democratised, recognised as being a product of ordinary people in everyday life. The accounts of these scholars of wit, punning, and poetic features in everyday conversation help to free our notions of creativity from associations with commercial value and stultifying connotations of the 'great man of genius', and to recognise instead the equal potential of, for example, marginalised groups in society. Such work, however, does not represent a wholesale rejection of this 'earlier' view of creativity. If it has rejected the focus of attention on the creative product, necessarily in lasting form, with qualities of commercial and scholarly value, it has however retained at the heart of the notion of creativity the concept of the individual acting with intent; what might be termed artistry. Innovative thinking is evident in so-called 'rule-breaking' as specific conventions, having been understood, are deliberately relaxed or broken for productive effect.

There is also in this focus on everyday creativity an increased recognition that creativity is often a collaborative enterprise. As Carter (2007: 598) proposes:

> Creativity emerges from interactional language encounters. It is often contingent and is produced and co-produced in domains in which new forms of social encounter . . . require the formation of different interactive practices and new identities.

This emphasis on the social element in the emergence of creativity is shared by Tusting and Papen (2008), whose work in the paradigm of the New Literacy Studies focuses on creativity as inherent in any activity of meaning-making. Rather than considering the text as detachable from its site of production and then amenable to being judged as 'more or less' creative, Tusting and Papen (2008: 6) examine the processes of text creation, as 'the sign-maker selects and combines the means and modes available to them to make meanings, a process that always entails adaptation and change' (Tusting and Papen, 2008: 7). They take as their task to examine local practices of selection and recontextualisation, explicitly avoiding an association of creativity with judgements of artistry as evident in the final product, while at the same time identifying 'the creativity inherent in seemingly mundane forms of written communication, even where the texts themselves might show little evidence of the language features normally associated with "artfulness"' (ibid.)

I do not, however, advocate a complete rejection of value-laden stylistic analysis. Instead I prefer to work with Maybin's (2006: 414) suggestion of a 'delicate balance' which avoids asserting either that creativity is an essential property of particular texts or that all meaning-making is creative, depending solely on the value judgements of the reader. Meaning-making is an interactive process; judgements of creativity are cultural acts; the reader is part of that culture and interaction, and so is the text.

This perspective, I believe, has the potential to open our eyes to creativity across many sites and practices where new challenges are encountered within the intrinsically imprecise, imperfect phenomena of social interaction (Rommetveit, 1992). The creative act may occur at the relatively macro scale, in for example the design of a space, or in a tiny piece of textual invention. One site at which this view of creativity was dramatically illustrated for me was the Schome Park project, an educational project for teenagers that I was fortunate enough to be involved in for over a year.

Schome Park

The Schome Park programme was set up under the auspices of a voluntary umbrella community of educationalists, young people, parents, teachers and anyone else who wanted to join online discussions about the future of education in the twenty-first century. This online community has since 2005 been led by Peter Twining of the Open University, UK. The word 'Schome' is derived from an early characterisation of the environment as 'not school, not home' which is not a simple rejection of both of those domains, but rather a suggestion that the firm boundaries between them need to be shaken. The aim of the project was to investigate and enact new models of education, providing participants with increased range, responsibility and control over their learning and greater opportunities for collaboration. In this context, technology was seen not only as a tool to support and extend existing practices but also as providing the potential to transform ways of representing the world and of supporting learning.

In 2006 the Schome community decided to explore the potential of virtual worlds, considering their capacity to act as spaces in which visions of future pedagogies could be built and experienced (Squire, 2006). With funding from a number of organisations at various stages of the project and a great deal of further voluntary input (see Gillen *et al.*, 2009; Sheehy, Ferguson and Clough, 2010; Twining, 2009 for more details), the Schome community decided to use Teen Second Life, the youth version of the virtual world Second Life.[1]

This made it necessary for the community participants themselves to design and enact all activities on the island (later two islands) which were the location for the project. Schome Park was the first Teen Second Life 'closed' educational project in Europe (i.e. with access restricted to project members). Although initially a few resources were 'imported' from elsewhere in Second Life, once the project began it was up to participants to design activities, establish ground rules and construct community practices and discourses.

Having downloaded the client application, via their 2D computer screens, participants interacted remotely, while physically located in homes, schools, workplaces and after-school clubs, mostly in the UK but also in the USA. Interaction in this environment is through an avatar. All movement and communicative practices are controlled by the individual who works with their avatar via their keyboard. The effects projected include movement by walking, running or flying, communicating through synchronous chat and various other modes of written communication, and building new objects on the island and making them do things. Perhaps the most important point to make about all these activities is that, apart from a small number of participants who had experience of Second Life, most participants including me had to learn all these practices from scratch.

It is beyond the scope of my work in this chapter to give an overview of the enormous diversity of activities that occurred on the island/s over the 13 months of the project; Twining and Footring (2010) give a good overview. In this chapter I will focus only on a very tiny part of the Schome Park project, specifically a number of activities connected with archaeology carried out during a latter phase of the project.

Towards a virtual literacy ethnography

Owing to the great diversity of participants, the distribution of their practices across time and space, and the length and open-ended nature of the project, it is difficult to fairly characterise or even sample in a representative way the tremendous range of activities and experiences.

My approach to studying this Second Life community is essentially ethnographic, despite the fact that all interactions were computer mediated with no physical connection with participants or project personnel in other spheres of life (see also Boellstorff, 2008). The challenge here, as expressed by Vannini (2009: 6), is to 'recenter ethnographic methodology in a way that is more consonant with the subject matter of material and technoculture research'. I term my developing approach to data collection and analysis in this project *virtual literacy ethnography* (Gillen, 2009). I emphasise literacy, recognising that, in line with the consistently negative representation of young people's new media language (Thurlow, 2007: 214), the diversity of literacy practices in virtual worlds is often overlooked. In part this may be owing to the emphasis on the visual in discourses about virtual worlds and perhaps Second Life in particular; it is, however, actually an environment that demands constant deployment of literacy skills.

My research approach concentrates on the literacy practices in and around the project, in so far as these are accessible to me. Sources of data include:

- chatlogs provided by staff members that comprise the texts of synchronous dialogues in which they participated inworld through their avatar projections as well as instant messages between participants sent or received by them at those times;
- forum data – i.e. texts of the asynchronous discussions that were held online during the course of the project. Since the forum could be accessed from any computer with an internet link at any time, this was actually the most popular communicative domain;
- wiki pages – i.e. collaboratively authored web pages, with texts and images, especially used to advertise forthcoming events and record the Schome Community's activities;
- field notes and images that I crafted whole participating inworld.

Central to my ethnographic observation was an emphasis on social practice as a creative, emergent phenomenon, in line with Tusting and Papen's (2008) notion of meaning-making as an ongoing and highly contingent process of creative appropriation and adaptation of existing practices and resources. Scollon (2001) proposes that, at the particular point in time when somebody deploys a communicative tool, they are negotiating their way through the understanding of social practices they possess, however emergently as a novice, enabled by the mediational means provided to them. Bauman (2010) similarly asserts that communicative practices are historical and experiential emergents. As a community we had to gradually learn, for example, that the most effective way of organising and advertising events was to list them all on an 'events page' in the wiki; during the early phases of the project considerable effort was wasted in advertising events on posters inworld, which were often not encountered by people through their avatars until the event was over. So we had to develop practices for these different sites of engagement, not as a matter of simply developing individualised operational skills but in collaboration, learning what worked best for what purpose, developing appropriate genres and registers and even shifting what we wanted to do. 'Tools', as Latour (2005: 143) reminds us, 'are never "mere" tools ready to be applied: they always modify the goals you had in mind'.

Archaeology, creativity and literacy in Schome Park

As an academic discipline that could be tackled within the Schome Park project without the weight of school-based curricula possibly acting to stifle innovation, archaeology was initially chosen as one of the themes the community set out to explore. In the first phase of the project archaeology was introduced under the leadership of an academic archaeologist (Greaves, 2007). After the professional academic archaeologist left the project, however, some students decided that they would like to bring their interests in archaeology and classical history together and convened Time Explorer meetings, mostly on Friday evenings; 17 were held in total. I attended most but not all of them. Figure 11.1 shows a snapshot of a meeting. On the occasion pictured here, the topic for discussion was the Caves of Lascaux. Participants interacted with websites external to the project while keeping Schome Park open in another window of their screen, discussing thoughts and opinions through their avatars.

I would like to argue here that the act of holding a meeting in Schome Park about archaeology was in itself a 'creative' act. I recognise that this claim depends largely upon the relative novelty of the event and indeed therefore that of the Schome Park project itself. As meetings about archaeology in virtual worlds become commonplace it will become increasingly difficult

FIGURE 11.1 Discussion of the Caves of Lascaux
(© Schome)

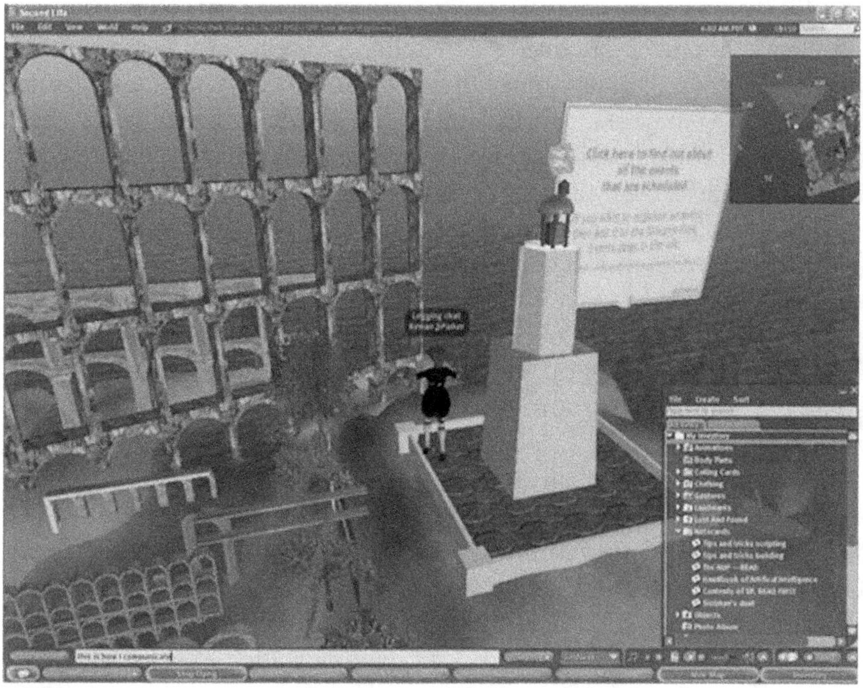

FIGURE 11.2 Screenshot of virtual construction site
(© Schome)

to make such a claim. At the time, however, the practices we worked out involving prior preparation of one or two people in the selection of external, web-based resources with the structuring of a discussion in a regular space and time felt at the very least a fresh way of engaging with issues of archaeology and classical history. Adopting Tusting and Papen's (2008) criteria for identifying the complexity of practices, omitting aesthetic judgements of the texts produced, the creativity of this activity becomes clearer.

Another kind of evidence to support this claim of creativity relates to the activity of construction in the virtual world. Figure 11.2 is a screenshot, which shows my avatar, Rowan, visiting the island and hovering close to a corner of it that had recently been used by the Time Explorers. They had been building Roman aqueducts and a monument. A poster board floating above the sea is also visible. It is important to note that this image is an imperfect rendering of the full colour screen; tones have been flattened and the sounds have obviously been removed. At least, however, this reductive image establishes the complexity of the demands on reading and writing present in this environment, even before interaction with others is considered, let alone any building activity. Rowan has just opened the menus she needs in order to navigate (see the minimap) on the top right of the screen, and she has her inventory open (bottom right) in order at this moment to locate a notecard, typically some kind of guide, created by one member in the community to be shared with others. The chat bar (bottom left) shows the space into which messages can be typed that then float into the air close to the avatar, readable by any other close to her, then either disappearing as ephemeral or captured if one participant has set their software to preserve their chatlogs. From this solitary image, it is impossible to know how many people had worked to recreate the aqueducts or what their purpose was. From my participation and community records I do know that several other recreations of archaeological sites and artifacts were made including the levels of Roman roads, 'El Castille' and the Mayan temple at Chichen Itza, Mexico.

Although I find the main structure of the aqueduct attractive, I would find it difficult to claim that the construction of the aqueduct visible in Figure 11.2 is necessarily a particularly creative act. Certainly some constructions in the virtual world did appear particularly innovative; in this case I know nothing about the history of its construction and can infer little about the resources and actors involved. I am unwilling to consider every construction in Schome Park as creative, or the meaning of that evaluative judgement dissolves.

I propose that the useful move to make in investigating creativity in (multimodal) literacy practices is identical to that which can be made in the study of archaeology itself. When originally beginning my engagement with archaeology I perceived its central aim as being the identification of cultural artefacts and settlements (and note Jones' 1997 proposal for a cyber-archaeology along those lines). I gradually came to perceive a fit of ideas developed about mediated action and sites of engagement (Scollon, 1998, 2001; Norris and Jones, 2005) with contemporary discourses and practices of archaeology (see e.g. Robertson, Seibert and Fernandez, 2005). Human practices are socially conceived, purposeful acts of sense-making that are enabled by the material affordances of the environment and cannot be understood without appreciation of the cultural values oriented to by the actors. A socio-historical approach to human activities necessitates taking not only a forensic approach to the study of any cultural artifact but an endeavour to appreciate the material, social and temporal influences that shape and are shaped by artifacts/texts.

Therefore, within a virtual literacy ethnography I want to use as many sources of evidence as I can not merely to analyse an artifact such as a single text, but to conduct a forensic investigation of some of the practices around its use, to enable me to point with more confidence to claims of innovation and indeed creativity. Discourse is mediated action and so it is sensible to centre analysis not on a single text or artifact but rather upon an event, a focus of social action. The incident I select to be at the core of my investigation here is chosen in part because it was a memorable point in my ethnographic engagement, an event I experienced and considered creative at the time. It was a literacy event as defined by Heath (1983: 386): 'any action sequence, involving one or more persons, in which the production and/or comprehension of print plays a role'.

The event in question occurred when I (through Rowan, obviously) attended an archaeology meeting on Friday evening concerned with the topic of Pompeii. During the meeting I was startled on suddenly experiencing a fall of what seemed to be rocks, realising with a shock that over a few seconds I was experiencing the sensation of a volcanic eruption. The short-lived yet intensive event cannot be transparently 'replayed' any more than any other experience in life; however, partial records are available. As with other notable events in the life of the community, it was recorded in the Schome wiki: see Figure 11.3.

FIGURE 11.3 Volcanic eruption recorded in the Schome wiki
(© Schome)

As the project developed, the wiki came to be used to summarise events of the week; the advantages of this medium for this purpose emerged gradually. It was persistent, readily organised and searchable, amenable to collective authoring, and good at hosting combinations of text with images. A wiki is a site of engagement for fostering the development of new literacies in its enabling of collaboration and creativity. As people come to the page and read it, especially if they were involved with the events depicted or have something to contribute, they can virtually instantaneously contribute either an addition, amendment or deletion. Distinctions between reading and writing, so essential to traditional literacies (de Certeau, 1984) become blurred in such practices as each person addresses the audience of which, at the same time, they are a part.

An element of the functionality of a wiki that can be useful to the researcher is its 'history' tab (as demonstrated in Gillen, 2009). Accessing this can preserve all the different versions of the page that have been made, giving the researcher the ability to work out who inserted the images, who wrote and altered the texts etc. Unfortunately in this case the wiki page was at some late stage moved in its entirety, obliterating this later history and so in this case, rather as an archaeologist studying an object which has been removed from its context without proper recording, I am considerably hampered in tracing the activities through which it was created. I do remember that I contributed the second (i.e. top right) image and its caption. I certainly did not have the presence of mind to take a snapshot as the rocks rained upon my head, but rather returned to them a little later to take the snapshot and later deployed 'poetic licence' to craft the caption. Until I returned to this wiki page I had certainly forgotten that we had originally met on the island's beach; usually Time Explorers met inside a room as shown in Figure 11.1 – that would, I suppose, have been less effective or perhaps more difficult for whoever reconstructed the eruption. Who was it that accomplished this feat?

In search of an answer to this question I investigated the project's other communication tools. Lemke's (2005) notion of chronotope invoking culturally typical movements and pacings across trajectories of activity is wholly applicable to the consideration of uses of the different project tools in terms of literacy practices. Just as the wiki became recognised as the best medium through which to construct records of notable events, so an important channel for planning was the forum. Easily accessible, and offering asynchronous discussions, organised in linear threads, it was well suited to this purpose. However, being publicly shared (among project participants), any final details of planning the eruption could not have been posted there or it would not have been a surprise. Nevertheless I could search the forum for early traces of any relevant planning activities. Tracing back I identified a message posted in the forum by Marsbar9 Schomer on April 20, 2008, as the first in what became a complex sequence of interactions as the event was planned. He started a new topic thread, locating it within the 'Time Explorers' strand, called it 'SPP Archaeology session 11' and proposed:

> Please sign up to the session – in this session we'll start on the recreation. Please remember this is at the earlier time of 7.15.

The first six words were hyperlinked to a 'sign up page' in the wiki and evidently in both the inexplicit mention of the 'recreation' and appeal to memory, this posting is 'a link in a very complexly organized chain of other utterances' (Bakhtin, 1986: 69).

This first message was replied to with enthusiasm just six minutes later by MissVibia Schomer, another student; these two became the leaders of the eventual event. Over the next few minutes a discussion began to evolve as to the main activity of the planned session. Although a bulletin board is asynchronous, this one included the facility of indicating whether participants who had posted a message that is being read are online; therefore the discussion evolved with turn-taking that is closer to conversational synchronicity than often is the case on such forums. Nevertheless, the chronotope of this forum discussion is different from that of oral discussions.

The first point to note in Figure 11.4 below is the economy of action employed by Marsbars9 Schomer realising that he wishes to respond to a point made nine minutes before. As his posting will not appear directly below KitKatKid Schomer's proposed 'recreation' he quickly copies and pastes the part of that posting he wants to refer to before posing his question. It is impossible to know at this later date whether KitKatKid Schomer was online at that point; most probably not, in that MissVibia jumps in with a quick explanation seemingly alluding to a plan to reconstruct part of the process of using logs to move stones used in Stonehenge. Marsbar9's response employs some of the jargon of the Teen Second Life environment. In this short example we see something of the inextricable, associated qualities of lexicon, grammar and compositional structure that are characteristic of what Bakhtin (1986: 60) calls speech genres, that 'reflect the specific conditions and goals of each such area [of human activity]'.

Subsequent to the postings shown in Figure 11.4 some became far longer and more detailed. A discussion evolved as to the mechanics involved in the proposal; it is difficult within the constraints of this chapter to properly explain the challenges involved in any such activity. Kitkatkid Schomer

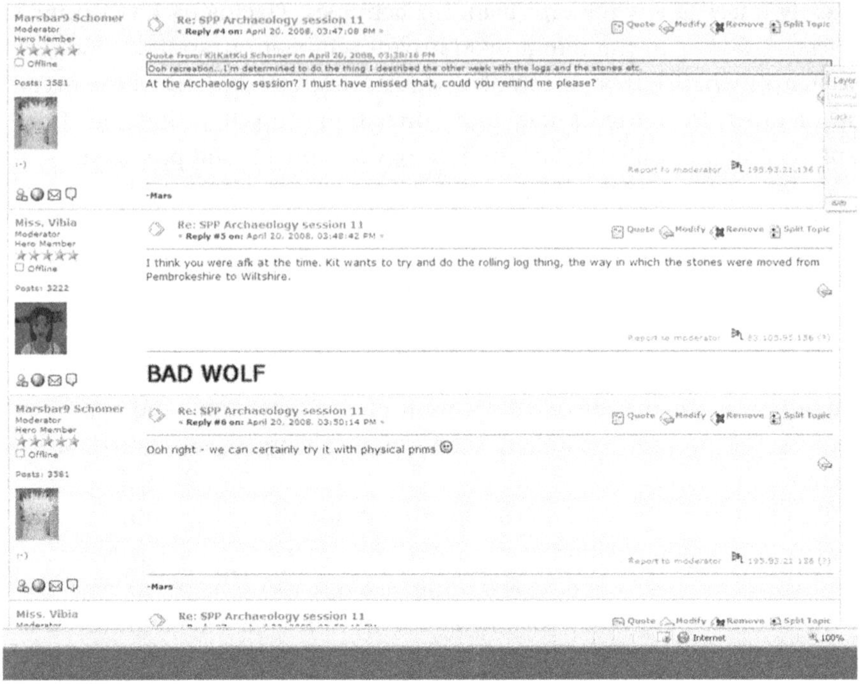

FIGURE 11.4 Forum discussion
(© Schome)

joked imaginatively that things could be moved uphill if 'we fatten ourselves up and push!'. Here he was alluding to the size of avatars as to a degree modifiable, knowing that his interlocutors would know this is irrelevant to processes of moving objects inworld.

The traces of planning the event then go cold; in order to keep the activities of building and scripting that must have been involved in creating the eruption secret, planning must have moved to more private channels, i.e. through intimate interactions inworld, undiscovered by others, and/or through instant messaging. I must mention that the amount of skill, knowledge and understanding required to create the eruption was considerable, demanding programming (hence literacy skills) in the environment far beyond my own capabilities.

The wiki events board gives a last-minute clue as to what must have been an essential element of the surprise; shortly before the meeting began, Marsbar9 posted an apology that he would be late and that the discussion about Pompeii should begin without him. This, I can now reconstruct, would have been to give himself the opportunity to hide out of sight in the sky, from where he must have launched the eruption.

The final source of record for the event is my own chatlog. My chatlog for the month of May 2008 is 63 pages in a single-spaced Word document and many other researchers contributed their logs. I have selected an extract from approximately three minutes' worth of activity (see Figure 11.5) beginning from the first verbal evidence of the eruption event.

```
[2008/05/09 11:48]  Eli SParker: OI !
[2008/05/09 11:48]  Vibia Schomer: Im interested in both the Geology and
Archaeology
[2008/05/09 11:48]  Vibia Schomer: =)
[2008/05/09 11:48]  You: ow ow ow
[2008/05/09 11:48]  marsbar9 Schomer: eek :P
[2008/05/09 11:48]  Vibia Schomer: AHHHHHHH
[2008/05/09 11:48]  You: OW OW OW
[2008/05/09 11:48]  Yati SParker: woouuuu
[2008/05/09 11:48]  Vibia Schomer: ERUPTION!
[2008/05/09 11:48]  Elsa SParker: OWWWW!
[2008/05/09 11:48]  marsbar9 Schomer: lol
[2008/05/09 11:49]  Elsa SParker: we need an emergency proceedure!
[2008/05/09 11:49]  Vibia Schomer: PYROCLASTIC FLOW!
[2008/05/09 11:49]  marsbar9 Schomer: :P
[2008/05/09 11:49]  Elsa SParker: LOL VIB!
[2008/05/09 11:49]  Vibia Schomer: (without the hot air) =p
[2008/05/09 11:49]  marsbar9 Schomer: in an emergency follow Elsa and dive
under the sand :P
[2008/05/09 11:49]  Vibia Schomer: =p
[2008/05/09 11:49]  Vibia Schomer: lol
[2008/05/09 11:49]  Yati SParker: lol sky has fallen down
[2008/05/09 11:49]  You: that was a huge surprise
[2008/05/09 11:49]  marsbar9 Schomer: =P
[2008/05/09 11:49]  Eli SParker: lol
[2008/05/09 11:49]  You: did someone get a frap?
[2008/05/09 11:49]  Vibia Schomer: Well, that is large pumice... =p
[2008/05/09 11:49]  marsbar9 Schomer: you mean machinima?
[2008/05/09 11:49]  marsbar9 Schomer: It happened so quick I doubt there
was chance :P
[2008/05/09 11:50]  Vibia Schomer: =p
[2008/05/09 11:50]  You: yes Vibia I had foirgotten it is pumice
[2008/05/09 11:50]  Vibia Schomer: =)
[2008/05/09 11:50]  Elsa SParker: gosh how exciting, I may need a lie down
[2008/05/09 11:50]  You: even sound effects!
[2008/05/09 11:50]  marsbar9 Schomer: lol
[2008/05/09 11:50]  Vibia Schomer: I'm sad, i remember these things! =)
[2008/05/09 11:50]  You: who did that it was amazing
[2008/05/09 11:50]  marsbar9 Schomer: it was me :P
[2008/05/09 11:51]  Rowan SParker is hugely impressed
[2008/05/09 11:51]  Fox Phlox: very impressive
[2008/05/09 11:51]  Elsa SParker: there's flowing lava on beta
[2008/05/09 11:51]  Yati SParker: that very coolll l
[2008/05/09 11:51]  You: that's true
[2008/05/09 11:51]  marsbar9 Schomer: ye - that is very good
[2008/05/09 11:51]  Fox Phlox: could we do a fall of ash somehow?
[2008/05/09 11:51]  Fox Phlox: maybe with a particle effect?
[2008/05/09 11:51]  Vibia Schomer: Particals
[2008/05/09 11:51]  Vibia Schomer: =)
[2008/05/09 11:51]  Elsa SParker: you should do it there Mars, teh
combination would be realyl cool
[2008/05/09 11:51]  marsbar9 Schomer: ye it would- would make a good photo
lol
[2008/05/09 11:51]  Fox Phlox: i don't know how to do particles
[2008/05/09 11:51]  Vibia Schomer: Neither do i... =p
[2008/05/09 11:51]  Vibia Schomer: Maaarrrrssss...... =p
[2008/05/09 11:52]  marsbar9 Schomer: Maaaaarrrsss scripting? :P
[2008/05/09 11:52]  marsbar9 Schomer: I could try with prims
[2008/05/09 11:52]  You: gosh even Mars can't do everything
[2008/05/09 11:52]  Fox Phlox: Eli?
[2008/05/09 11:52]  You: as far as I know....
[2008/05/09 11:52]  marsbar9 Schomer: indeed rowan lol
```

FIGURE 11.5 Chatlog from the eruption event

Owing to the software conventions, the name of the avatar producing the turn appears immediately to the left of the colon. 'You' means Rowan; anybody with the surname Schomer is a student and SParker an adult.

Reading the transcript is likely to be a little unsettling for the linguist used to analysing written language or transcripts of spoken language. Spoken language, even when mediated through technologies such as the telephone, is characterised by fine turn-taking. Close attention by linguists has revealed overlap to be relatively rare; and conversation analysis attuned scholars to its significance (Sacks, Schegloff and Jefferson, 1974). Written language is most often asynchronous, although the bulletin board, as discussed above, may well entail close proximity in time in the output of turns. In chatlog data such as that shown in Figure 11.5, the turns appear sequential but actually this conceals considerable overlap in terms of their production. Some participants in the Schome Park project found the intensity of the pace and necessity, or at least desirability, of constantly attending to new items presented for reading at the same time as one is composing a response to an earlier turn stressful. As Hutchins (1998: 239) observes:

> Meanings seem to be in the messages only when the structures with which the message must be brought into coordination are already reliably in place and taken for granted. The illusion of meaning in the message is a hard-won social and cultural accomplishment.

Although it appears second in the transcript, it is evident to the practised reader that Vibia's second turn is much shorter than the first, and therefore (almost certainly) the writing of it would have begun before Eli's. Therefore it is a response to an earlier turn, indeed part of a more general discussion about Pompeii, couched indeed in more standard sentences. This exchange, although for the most part concerned with a sudden orientation to the salient event, of the simulated lava flow, features considerable intertextuality; for example, marsbar9's proposed 'emergency procedure' – 'follow Elsa and dive under the sand' – refers back to a conversation two minutes earlier when Elsa (the avatar, obviously) was 'found' almost under a rock, half-hidden in the sand. Indeed Elsa's later comment 'gosh how exciting, I may need a lie down' seems to me to cleverly index not only a social practice of embodied reaction to a dramatic event in one's personal life, thus conveying appreciation of the emotional power of the eruption, but also continuing the playful discursive theme of her avatar as prone. This, to me, is at the same time both witty and an indication of the liminality that the experience of interacting in a virtual world can bring, at least during such immersive events.

In the heat of the moment, evidently a number of participants were faced with the immediate desire to convey expressions of surprise that would in 'real' life be couched in an oral mode. Reactions of surprise to a sudden event are not usually in the moment conveyed through written means but of course, as already explained, the only channel of communication open to us was a written one. So there are a number of attempts to express orthographically nonverbal explanations like the initial 'OI!' but, I would suggest, convey a positive evaluation through the crafting of some difference in form. I recall how weak at the time I felt my 'ow ow ow' to be, yet, stunned, I could not immediately think of anything more expressive to do than repeat it in more emphatic capitals. Vibia, to whom the event did not come as a surprise, had evidently already planned an exclamation that encapsulated one of her pedagogic intentions as a leader of the event; she moved first into a more semantically meaningful exclamation: 'ERUPTION!' and then into her key terminological teaching point: 'PYROCLASTIC FLOW!' The nature of this and how it can be further developed in simulations in Schome Park became the shared focus for subsequent interactions. What inspirational teaching! (I feel I ought to mention again in case it is necessary that Mars and Vibia are teenagers in actuality; the other participants in this extract are adults.)

Drawing on work by Gee (1999), Lankshear and Knobel (2006: 71–2) write:

> literacies are always about much more, and involve much more, than just the production of texts. They are (also) contexts or pretexts for enacting and refining memberships of Discourses that include such dimensions as feeding back, providing support, sharing knowledge and expertise, explaining rules, sharing jokes . . . enacting an affinity.

In my view this event dramatically portrays the rich textures of the 'economies of interaction' (Jones, 2010) and contribution towards positive affect that the literacy practices of the SPP did at times instantiate within the community. The eruption, while being a literacy event, was one of my most vivid experiences in any formal or informal learning situation I have encountered. In virtual worlds one has the potential not just to undergo a sensation but to simultaneously interact with it and with the people who are also engaging with the environment. The next person to come along, whether to the beach inworld, the forum or the wiki, will find a changed environment, according to whatever action he or she takes. Here we see, rather than the presumption of a shared text resource for reading: 'radically changing forms and functions of texts, which go beyond traditional conceptions of what literacy is and has been' (Kress, 2010: 6).

I considered then, as I do now, the eruption to have been a creative and effective learning experience, creativity not in the 'great men of genius' style, but in the refashioning of available (multimodal) resources in startling combination to a wonderful new effect (Gillen, 2007; Willis, 2009).

Conclusion

I have presented here my own judgements in discerning creativity in thematically linked acts of mediated discourse on different timescales (Lemke, 2005). I have endeavoured to present evidence to illustrate how the following may be considered creative:

- Schome Park itself as a 'designed experience';
- the Time Explorers' practices of studying archaeology in a virtual world;
- the volcanic eruption during the discussion of Pompeii;
- aspects of the texts produced around that eruption.

At the same time as organising these claims through this temporal scale I resist any causal link between them, especially as might be deemed technologically determined. The use of a virtual world is not necessarily any more innovative or creative than that of a design of a school or other community space might be; it was the combination of the imaginative, purposeful group of people acting in this nexus of practice (Scollon, 2001) that made it so at some points.

Mediated actions, including those that involve literacy, involve temporality and spatiality, are distributed in environments that themselves bear the marks of former human cultural activity, and are social practices imbued by the habitus of individuals (Scollon, 2001; Gillen and Cameron, 2010). An overarching emphasis on the dialectical nature of activity systems is characteristic of cultural historical activity theory as defined by Stetsenko, (2009: 126):

> According to this vision, human development is rooted in, derivative of, instrumental in, and constituted by the material collaborative social practices of people (i.e., human goal-directed, purposeful, collaborative activities) aimed at transforming their world.

This serves as a superb perspective on not only the experiences of Schome Park as reflected here but also encapsulates the framework for approaching literacy and archaeology I have put forward. In both domains we can see the overlapping timescales of simultaneously present rhythms of human activity in complex ecosocial systems (Lemke, 2000). Finally, then, in this chapter I have tried to respond creatively myself in considering archaeology as topic and methodological inspiration entwined with mediated discourse analysis.

Note

[1] Teen Second Life and second Life are trademarks of Linden Research, Inc. Teen Second Life has ceased to operate since the time of the project reported on here.

References

Bakhtin, M. M. (1986) *Speech genres and other late essays*. Austin: University of Texas Press.

Bauman, R. (2010) '"It's not a telescope, it's a telephone": Encounters with the telephone in early commercial sound recordings' in *Language ideologies and media discourse: Texts, practices, politics*, Johnson, S. and Milani, T. (eds). London: Continuum, 252–76.

Boellstorff, T. (2008) *Coming of age in second Life*. Princeton: Princeton University Press.

Carter, R. (2004) *Language and creativity: The art of common talk*. London: Routledge.

Carter, R. (2007) 'Response to special Issue of Applied Linguistics devoted to language creativity in everyday contexts'. *Applied Linguistics* **28** (4) 597–608.

Cook, G. (2000) *Language play, language learning*. Oxford: Oxford University Press.

de Certeau, M. (1984) *The practice of everyday life*, (trans. Rendall, S. F.). Berkeley: University of California Press.

Gee, J. P. (1999) *An introduction to discourse analysis: Theory and method*. London: Routledge.

Gillen, J. (2007) 'Derwent's doors: Creative acts'. *Mind, Culture and Activity* **14** (3) 150–9.

Gillen, J. (2009) 'Literacy practices in Schome Park: A virtual literacy ethnography'. *Journal of Research in Reading* **32** (1) 57–74.

Gillen, J. and Cameron, C. A. (eds) (2010) *International perspectives on early childhood research: A day in the life*. Basingstoke: Palgrave Macmillan.

Gillen, J., Twining, P., Ferguson, R., Butters, O., Clough, G., Gaved, M., Peachey, A., Seamans, D. and Sheehy, K. (2009) 'A learning community for teens on a virtual island – The Schome Park teen Second Life pilot project'. *eLearning Papers* **15** The New Learning Generation. Retrieved 15 July 2010 from http://www.elearningpapers.eu/index.php?page=doc&doc_id=14397&doclng=6

Greaves, A. (2007) 'Reconstructing Hadrian's Wall in Second Life'. *Higher Education Academy: History, Classics and Archaeology Case Studies*, Retrieved 22 February 2010 from http://www.heacademy.ac.uk/hca/resources/detail/reconstructing_hadrians_wall

Heath, S. B. (1983) *Ways with words*. London: Cambridge University Press.

Hutchins, E. (1998) *Cognition in the wild*. Cambridge, MA: MIT Press.

Jones, Q. (1997) 'Virtual-communities, Virtual settlements and Cyber-archaeology: A theoretical outline'. *Journal of Computer-Mediated Communication* **3** (3).

Jones, R. H. (2010) 'Technology and sites of display' in *The Routledge handbook of multimodal analysis*, Jewitt, C. (ed.). London: Routledge, 114–26.

Kress, G. (2010) 'The profound shift of digital literacies' in *Digital literacies*, Gillen, J. and Barton, D. (eds), Research briefing for the TLRP-TEL (Teaching and Learning Research Programme – Technology Enhanced Learning). London: London Knowledge Lab, Institute of Education, 6–7. Retrieved 22 February 2010 from http://www.tlrp.org/docs/DigitalLiteracies.pdf

Lankshear, C. and Knobel, M. (2006) *New literacies: Everyday practices and classroom learning* (2nd edn). Maidenhead: Open University Press.

Latour, B. (2005) *Reassembling the social: An introduction to actor-network-theory*. Oxford: Oxford University Press.

Lemke, J. (2000) 'Across the scales of time: Artifacts, activities, and meanings in ecosocial systems'. *Mind, Culture and Activity* **7** (4) 273–90.

Lemke, J. (2005) 'Place, pace, and meaning: Multimedia chronotopes' in *Discourse in action: Introducing mediated discourse analysis*, Norris, S. and Jones, R. (eds). London: Routledge, 110–22.

Maybin, J. (2006) 'Locating creativity in texts and practices' in *The art of English: Everyday creativity*. Maybin, J. and Swann, J. (eds). Basingstoke: Palgrave Macmillan, 413–34.

Maybin, J. and Swann, J. (2006) *The art of English: Everyday creativity*. Basingstoke: Palgrave Macmillan.

Norris, S. and Jones, R. H. (2005) *Discourse in action: Introducing mediated discourse analysis*. London: Routledge.

Robertson, E., Seibert, J. and Fernandez, D. (2005) *Space and spatial analysis in archaeology*. Calgary: University of Calgary Press.

Rommetveit, R. (1992) 'Outlines of a dialogically based social-cognitive approach to human cognition and communication' in *The dialogical alternative: Towards a theory of language and mind*, Wold, A. H. (ed.). Oxford: Oxford University Press, 19–44.

Sacks, H., Schegloff, E. and Jefferson, G. (1974) 'A simplest systematics for the organisation of turn-taking for conversation'. *Language* **50** 696–735.

Scollon, R. (1998) *Mediated discourse as social interaction: A study of news discourse*. London and New York: Longman.

Scollon, R. (2001) *Mediated discourse: The nexus of practice*. London: Routledge.

Sheehy, K., Ferguson, R. and Clough, G. (eds) (2010) *Virtual worlds: Controversies at the frontier of education*. Hauppauge, NY: Nova.

Squire, K. (2006) 'From content to context: Videogames as designed experience'. *Educational Researcher* **35** (8) 19–29.

Stetsenko, A. (2009) 'Vygotsky and the conceptual revolution in developmental sciences: Towards a unified (non-additive) account of human development' in *World Yearbook of Education 2009. Childhood Studies and the Impact of Globalization: Policies and Practices at Global and Local Levels*, Fleer, M., Hedegaard, M. and Tudge, J. (eds). New York: Routledge, 125–41.

Thurlow, C. (2007) 'Fabricating youth: New-media discourse and the technologization of young people' in *Language in the media: Representations, identities, ideologies*, Johnson, S. and Ensslin, A. (eds). London: Continuum, 213–33.

Tusting, K. and Papen, U. (2008) 'Creativity in everyday literacy practices: The contribution of an ethnographic approach'. *Literacy and Numeracy Studies* **16** (1) 5–23.

Twining, P. (2009) 'Exploring the educational potential of virtual worlds: Some reflections from the SPP'. *British Journal of Educational Technology* **40** (3) 496–514.

Twining, P. and Footring, S. (2010) 'The Schome Park programme: Exploring educational alternatives' in *Researching learning in virtual worlds*, Peachey, A., Gillen, J., Livingstone, D. and Smith-Robbins, S. (eds). London: Springer, 53–74.

Vannini, P. (ed.) (2009) *Material culture and technology in everyday life: Ethnographic approaches*. New York: Peter Lang.

Willis, P. (2009) 'The accidental ethnographer and the accidental commodity'. *Ethnography* **10** (3) 347–58.

CHAPTER 12

Designing location-based mobile fiction: the case of *NarraHand*

Andrew Morrison, Henry Mainsah, Idunn Sem and Martin Havnør

Contexts for discursive design

The snow is almost a metre thick. Eric Kigundu is walking fast to keep warm. He takes off his glove and digs around in his jacket pocket for his mobile phone, thinks of his visit home to Uganda. He walks a few more paces and stops. He breathes in slowly, and then lets himself fall into the experimental character he's chosen. Preacherman. He opens the NarraHand app. Selects his fictional name. Chooses to make a new entry – and begins to write.

Eric smiles to himself, knowing the entry will be logged via GPS to his location on the snowy street but that his text is about a clammy evening in Kampala. Re-reading his entry, he saves it, takes a photo of an advert for a cheap phone card in the shop window beside him, and uploads these to the story space. His hand's now bonily cold, so he pockets the phone and retrieves his glove, laughing at the split between time, place and space he's just created. 'That should give the other authors something to think about'.

Locative media has often taken the form of functionalist positioning products and services which, for example, help users identify the nearest underground station. Alongside these, however, other more playful and artistic approaches to mobile communication on city streets have developed, such as urban gaming. In this chapter we present reflections on the design and trialling of a mobile application for collaborative fiction. As the opening scenario illustrates, this design is part of a wider research project into genre, mobile media and communication in which mobile media, software, mapping, experimental use and exploratory expression have all been included. The application we have developed, called *NarraHand*, uses the software JAVA and is based on Global Positioning Services (GPS). This locative technology is used as design material and as a semiotic resource for the shaping of an eventual mobile, character-led fictional work within a core area of a capital city in Scandinavia intended to be used by adult African immigrants as part of a project on the expression of diasporal identity via mobile, social media.

This chapter presents an ethnographic study of iterative, participatory co-design. In it we will present several narrative vignettes and link these to emerging research on mobility, place and space that are now beginning to be taken up at the level of their mediated design (Løvlie, 2009; Messeter, 2009; Nova and Girardin, 2009). In doing so we address two questions: 1) How can communicative tools be designed as resources to enable collaborative creativity? and 2) How can the design of this tool itself be seen as a collaborative creative process?

In several respects, this design experimentation and participatory research may be called 'discursive design' (Arnall and Martinussen, 2010). We prefer the term discursive design to the often more technologically driven 'interaction design' because it focuses on the creation of tools and means for mediated communication, taking into account existing discursive practices and their impact on the design. Further, discursive design operates as a heuristic device: it helps us think about digital materials in terms of their potential for discursive enactment and engagement. A discursive design view is oriented beyond seeing design as simply providing solutions to problems to seeing it as designing for shared meaning-making. This is an important concept for studies of discourse and creativity where, in analysis of discourse products and processes, design practices and design research are largely invisible. Discursive design needs to be included in studies of discourse and creativity where the design of communication technologies can have considerable influence on discourse practices.

In contrast to many studies of design, we take a developmental and collaborative view of both discourse and creativity. In focusing on a shared developmental process, we reflect on multiple design activities and mediations with respect to what is in effect collaborative 'composition' (Morrison, 2010). This means that, as the project progressed, we needed to find a fit not just between technical demands and the communicative intentions of potential users, but also between the contributions of different team members with different backgrounds and competencies. This team included: a Hallidayian-schooled discourse analyst working in new media, narrative and design research; a media researcher in the area of youth, social media and identity; a media designer; an informatics developer-researcher from a university technology lab along with its programmers, and the support of a leading commercial mobile phone services programmer. The development team also worked in various iterative design trials with an experimental group of six users from four African countries.

This chapter covers only the first phase of the larger developmental design research project. Importantly, this first phase centred on the creative, collaborative and developmental process of design. The second phase, already under way, concerns the actual articulation of a character-led collaborative mobile, locative fiction that is being created with the tools developed in the first phase.

Discourse and design

Henry's writing up his PhD dissertation in the depths of winter in Oslo. Meanwhile, Ole and Andrew are at the other end of the earth. They're visiting the University of Cape Town on an educational research project into motivating and supporting the uptake of media technologies by humanities educators. It's a quiet hot March night. A slight breeze rustles the oak leaves outside.

'I was re-reading the Bødker and Børg Andersen', says Ole. 'You know, the multiple activities, multiple mediation article'.

Andrew laughs. 'Funny you mention it.' He pauses then asks, 'Do you think, Ole, we could make a mobile narrative interface, a locative media storyspace, a collaborative platform?'

'Sure', says Ole, not missing a beat. 'We'd use JAVA . . . I know probably the best company to help us do that.'

Through the window they see that the moon has come up. In the distance the mountains are outlined.

In adopting a research-by-design approach to mobile collaborative tools and spaces, our motivation has been to engage in communication design for mobile fictional discursive expression. In such a project, Hayles (2008: 157) reminds us that:

> . . . computation ceases to be a technical practice best left to software engineers and computer scientists and instead becomes a partner in the co-evolving dynamics through which artists and programmers, users and players, continue to explore and experience the intermediating dynamics that let us understand who we have been, who we are, and who we might become.

A design view of discourse, creativity and technology offers some means for engaging with given techno-mediational conventions and practices while opening out to innovative ways of shaping digitally mediated communication. This is territory that needs to be investigated discursively in the activities of designing.

Our main claim is that design is an under-articulated aspect of multimodal (Kress and van Leeuwen, 2001) and multimediational discourse (Morrison, 2010). Although studies have recently appeared which address multimodal discourse and narrative (Page, 2010) and multimodality and mobility (Adami and Kress, 2009), these do not specifically address the *design* of multimodal-constituted communication tools nor the demands or potential that emerging technologies present us with in terms of mediated communication. In *NarraHand* we focus on the design of communicative affordances and affordances for communication. Such a discursive take on design as activity and discourse as performative action builds upon work previously carried out in the co-design of a social calendaring service, *Underskog* ('Undergrowth', in Norwegian; Morrison *et al.*, 2010).

Our overall focus is on design for communication, broadly framed within an approach to discourse as mediated action (Norris and Jones, 2005). This approach focuses on discourse as a matter of actions performed via mediational means. Yet, in this framework, little attention has been paid to the design of mediational means themselves. Our focus on design for communication refers to ways in which contemporary mediational technologies may be conceptualised and iteratively implemented. The aim of this is to provide subsequent participants in digital discursive production with communicative affordances and semiotic resources that facilitate their personally crafted collaborative narrative expression. Without closer attention to the communication design of mediational technologies, more extensive understanding of collaborative and distributed features and processes of 'digital discourse' may be restricted.

Following Vygotsky (1962), our research is situated in a sociocultural framing of mediational technology and design. In this view, relations between tools and signs are realised via social and cultural practices and significations, not merely technologies nor single modes of discourse. Such articulations involve multiple activities and multiple mediations that are cast in a dialogical view of mediated communication (Shotter and Billig, 1998), not one that is uni-directional (a sender–receiver model). Our dialogical approach is also connected to studies that focus on creativity as collaboratively crafted rather than individually fashioned (Sawyer, 2006).

Here design, creativity and collaboration are also developmental in character (Moran and John-Steiner, 2003).

In design research the notion of affordances, drawn from Gibson (1979), is often cited to refer to the potential properties of materials and perceptions that are needed to realise the designed. We extend these views into a sociocultural perspective on designing, making use of the notion of the *communicative prototype* developed by Knutsen and Morrison (2010) in the design of mobile social software tailored for the upload and discovery of music by independent, non-commercial actors. Much has been made of the participatory and performative aspects of social media in the contexts of Web 2.0. 'Social media' and 'social software', however, are rarely researched communicatively in terms of their *designs for discursive enactment*, the constraints and affordances they entail and the dynamics of their critical and reflexive review and redesign via experimental and later wider public use. In this chapter, we argue that the concept and embodiment of *communicative affordances* are central to digital design with emerging social media and mobile technologies. The role of mediation (Wertsch, 1991) and the agency of 'prosumers' (Jenkins, 2006) in social media need to be more fully connected to creativity in design, discourse and technology, especially the burgeoning mobile market and related services.

In its conceptualisation, *NarraHand* aimed to encompass not only mixed media and mixed identities but also their instances as parts of mixed reality lifeworlds. The point of our co-design, in other words, was to devise a mediated mobile means of articulating links between media and identity. To do this, we decided to focus on the experiences of African immigrants in Europe, examining the potential that new communication technologies offer for the shaping of 'migration' narratives among a self-forming group of African 'authors' motivated to find ways to reflect on and share their experiences of home and diaspora (Mainsah, 2009). This focus draws on earlier work by the authors, especially Morrison's (2003) development and implementation of a stand-alone hyperfictional platform for use with students in Zimbabwe in the mid-1990s as a topological means of examining land, identity and voice. Inspiration was also taken from other innovations in digital locative narrative, such as the Yellow Arrow project (http://yellowarrow.net/v3/index.php).

Digital narrativity and design exemplars

Today, mobile social media are being taken up in many diverse global contexts. One example is the *m4Lit* project[1] in South Africa (http://m4lit.wordpress.com), which has investigated urban teens' use of mobile social networks as part of collaborative literacies development (Walton, 2009). For such users, GPS-enabled mobile phones are often the primary points of access to the internet. Such a project highlights the self-authoring aspects of mobile communication and its purposive design for potential engagement and self-generated content.

Locative media works are still in their infancy, as is research on their design and analysis, much of which is game related (e.g. Diamantaki *et al.*, 2009). *34 North 118 West* (http://34n118w.net/), for example, centres on GPS and tablet computing to explore the early industrial period of Los Angeles. *Mr Beller's Neighbourhood*[2] is an online collaborative and locative work that gathers largely non-fiction stories in New York city and allows access via a map interface or website, also thematically aggregating these over time. Rettberg (2005) is one of few digital narrative scholar-practitioners who tackles *collaborative digital fictional* narrative both as an author and as an analyst. He discusses processes of co-writing a hyperlinear, playful and picaresque hypernarrative, *The Unknown*, over four years. In this work, geographical space rather than chronology structured the story processes.

As locative media, *Blast Theory* (http://www.blastheory.org) takes mobile and locative technologies and transforms them into participative urban events situated in real city spaces (Benford *et al.*, 2006). In this way, discursive agency and participation are realised via mobile media technologies within and across cityscapes. *The 21 Steps* by Charles Cumming,[3] part of a set of web-based stories by Penguin, also does this by providing readers with a map interface in which mobility paths of characters across large urban and geographical spaces are effectively animated. The reader, however, is static and bound to a desktop or laptop.

NarraHand is situated in an interactive fictional and digital collaborative narrative 'cast' that differs from classical narratology in that it is, as Ryan (2006: xxi) argues, 'simulative rather than representational, emergent rather than scripted, participatory rather than receptive, and simultaneous rather than retrospective'. In exploring these characteristics, at a design level, we needed to further investigate a mix of modes, mind processing and semiotic channels. In addition to distinguishing between immersive and interactive aspects of interactive narrative, Ryan also argues we need to look at narratives that transverse media. *NarraHand* aimed to do so by including text and image, geotagging, 'labelling' and proximity and identity markings.

Ryan (2007) further argues that there is a 'split condition of digital textuality' into interactive fiction or gaming. She envisages a middle ground open that now needs to factor in mobile media and multimodal design as part of what she distinguishes as *an opportunity for free play* rather than playing to win (see also de Lange, 2009). On *NarraHand*, this potential for free play, however, demanded we design a system for potential engagement, first, for collaborative composers of mobile narratives and, second, for readers. In both design and analysis of mobile narrative, further links remain to be made between developments in urban gaming and urban computing (e.g. Mäyrä and Lankoski, 2009) and the existing body of desktop and standalone interactive fiction.

We see exploratory, discursive knowledge-building in the domain of mobile media as part of a wider scope of 'communicative mobilities' (Urry, 2004). Engaging actively with the materialities of an emerging technology is central. In *NarraHand* this has moved from paper sketching and oral discussions to programming and media development in partnership between an experimental research lab setting and a mobile developer stretching its commercial map services to a multimodal narrative project. The result is a fully functioning application – with an inscribed set of communicative affordances – that ran first on contemporary JAVA-enabled devices and subsequently the iPhone. *NarraHand* is a mix of commercially served map tools for mobile phones with open source knowledge-management software crafted for collaborative narrative composition. A 'story' section is linked to a wiki (using *Confluence*) that allows for critical reflection, future personal narratives by readers and open discussion. In the next section we provide more details of the design of *NarraHand*.

Towards mediated design: *NarraHand*

Idunn and Andrew are huddled beside the radiator discussing narrative. As in earlier collaborative design research projects, a tension has arisen. Programmers and developers need to focus on their specialist components and artists theirs. In joint discussions on NarraHand we often end up focusing on technology.

Idunn and Andrew value the creative solutions arrived at by and with their computing colleagues. Now they need to better clarify designing communicative affordances for narrative expression.

'Let's go back to proximity', says Idunn. 'Would the narrative need both authors and readers to see how close a character is to another one? This isn't common in narrative. We can do that on the mobile map for bus stops, so . . .'

NarraHand was developed to explore the potential of handheld fictional communication in the context of personal and ubiquitous computing and social media as part of 'everyday life' (Boyd and Ellison, 2007; Manovich, 2009). Our goal was to develop an application that would give immigrant fictional storytellers a means to convey aspects of identity, place and space.

Designing affordances for discourse production in small, mobile screen-spaces is a difficult task for programming and information architecture as well for graphic, interaction and mediational design. In *NarraHand* we developed a number of main communicative affordances that are linked to meta-discursive narrative dialogical activities. These are embodied in a main menu. They cover (a) location, (b) recency, (c) identity, (d) relation, (e) proximity, (f) theme, and (g) comment.

NarraHand was designed to facilitate narrative and locative multimodal discourse production for 'authors' as well as multimodal, mobile consumption and commenting by 'readers'. Readerly-writerly dynamics are part of a developmental and multiliteracies view of the overall co-designing and trialling-in-use perspective.

In interdisciplinary projects different views need to have space and yet cohere. Potential contradictions and tensions may arise in the negotiation of different project partner competencies and perspectives. Tensions such as those illustrated above about keeping a narrative motivation to the fore need to be negotiated and discussed for communicative affordances to be realised. Importantly for mobile design, limited screen space and speed of access are critical factors; so too are relationships between discourse and creativity at the level of narrative tools and wider dynamics of contextual design. We tackled this by developing two main sections in the splash screen of *Narrahand*. This was done for the Nokia N95 and subsequently the iPhone, as illustrated here.

On the splash screen, the artistic, cultural construction is labelled 'Story' and is accompanied by a link to a wiki-based website called 'Context' (Figures 12.1 and 12.2). This second link offers background, related works and studies and experimental mobile research papers (some written on the device, designed to be read on the small screen or projected). Users of *NarraHand* would thus be able to toggle between the 'Story' and 'Context'.

Both 'Story' and 'Context' modes are further designed to be contextualised in a process of workshopping a full narrative piece with connections to the community of Africans in the city as well as other interests and debates concerning immigration, representation and identity.

A number of specific locative narrative affordances were developed. These relate to composing, reading and commenting. These needed to be crisply shown on the small screen and easily traceable in a pop-up menus. A navigation and zoom bar was provided for rapid shifting across the environment and magnifying specific locations. The core 'Map' interface with Points of Interest (POIs) marked by character can be seen in different modes (Figures 12.3 and 12.4). It is possible to toggle between these views while in the application when engaged in narrative composition, wayfinding or reading.

In the development of such affordances, having a working application in our hands was important for giving us actual experience with which to refine that application. However, this too would mean negotiations were needed between balancing narrative needs and technical possibilities. These are taken up in the next section.

Designs in our hands

It feels like an afternoon off but it's just an excuse to play! Idunn and the guys from InterMediaLab are off into the city to test NarraHand. Jeremy, Thomas and Ole have signed up with stand-in characters and logos. Idunn's arranged this and named her character Per Spook (a Norwegian fashion designer). Today Martin's online to make adjustments as they go. The 'authors' set off to walk, cycle and pause in spots around the suburb where Andrew lives. Idunn tries to provoke Thomas to pass nearby her and comment on her character's looks; she then alters another character's text to see how this feels as a function.

Andrew (project leader) is keeping some critical distance and is at home preparing dinner for the group.

As is often her role in the lab, Idunn is messaging and ringing the team while making her own entries.

We've been eating a while when Martin arrives from work. He's beaming. 'You guys, it works!' Now the discussion starts. Ole starts: 'We could change the . . .'

In developing affordances for narrative mobile expression, one challenge is how to allow for multiple perspectives and voices but also when to shift from experimental iterations to design decisions and final implementation processes. Designing for mobile screens is beset by spatial constraints, and not a task easily conducted by interaction designers more used to the expanse of the laptop or desktop.

Largely because of cost and download rate, we chose to concentrate on written words and still images. After testing we decided on a font, colour palette and size that would be readable on a medium-sized handset and outdoors in average sunlight. We also had to make this aesthetic and functionalist decision on the basis of selecting one leading mobile phone early in the project that had a stable GPS capacity, the Nokia N95. While having to contend with constraints, such as a system font that was fixed, we needed to design an interface that would also scale relatively reliably to other

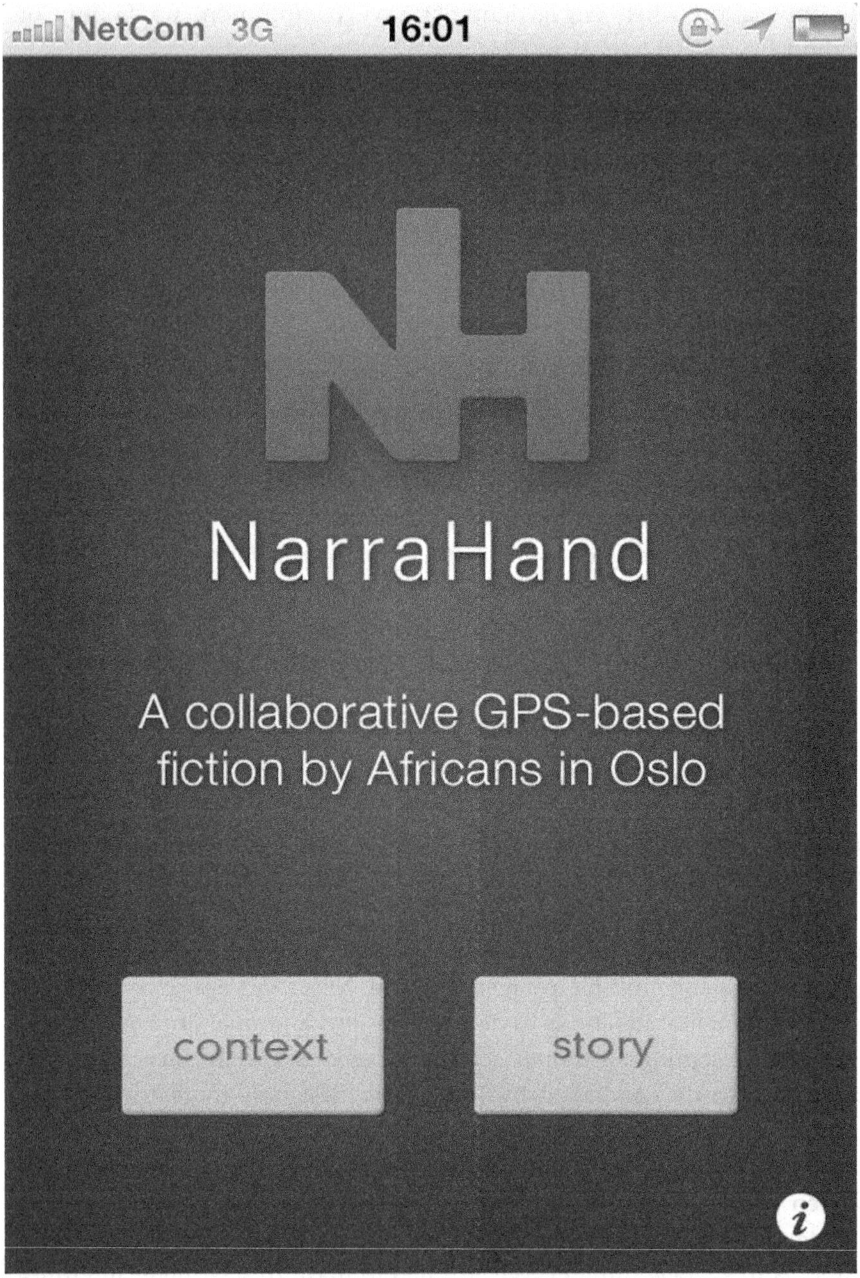

FIGURE 12.1 The splash screen for *NarraHand* with options 'Story' and 'Context' (© Andrew Morrison)

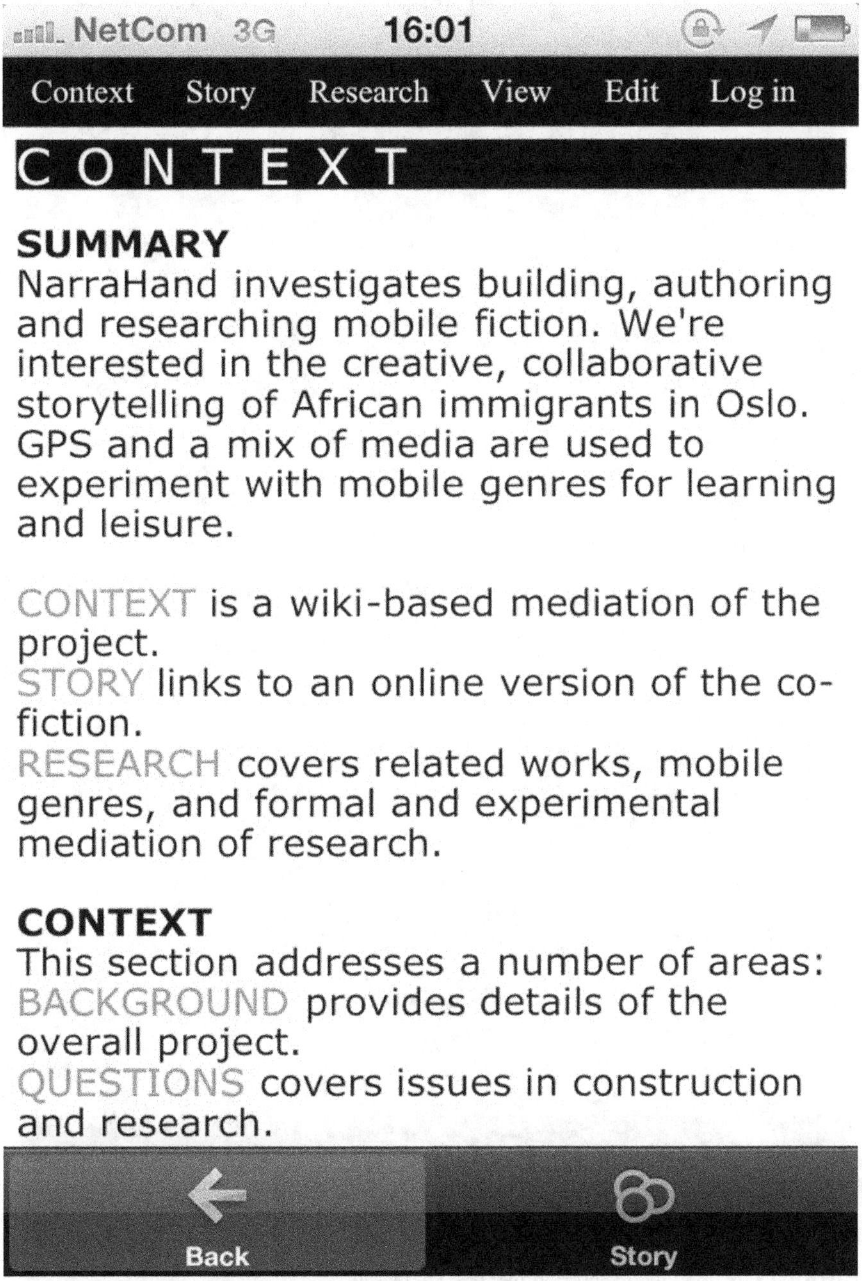

FIGURE 12.2 'Context' provides access to background, blog and research
(© Andrew Morrison)

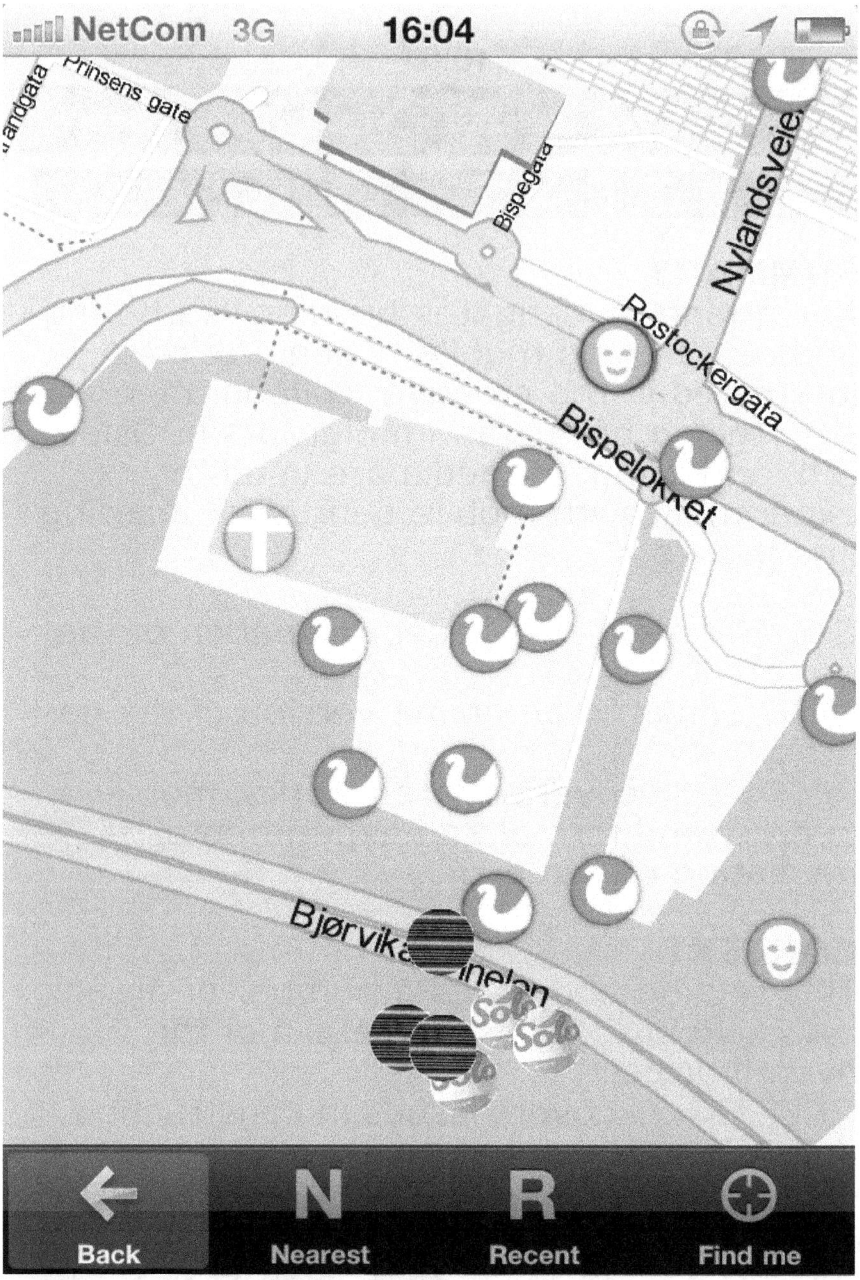

FIGURE 12.3 Map mode showing five test characters
(© Andrew Morrison)

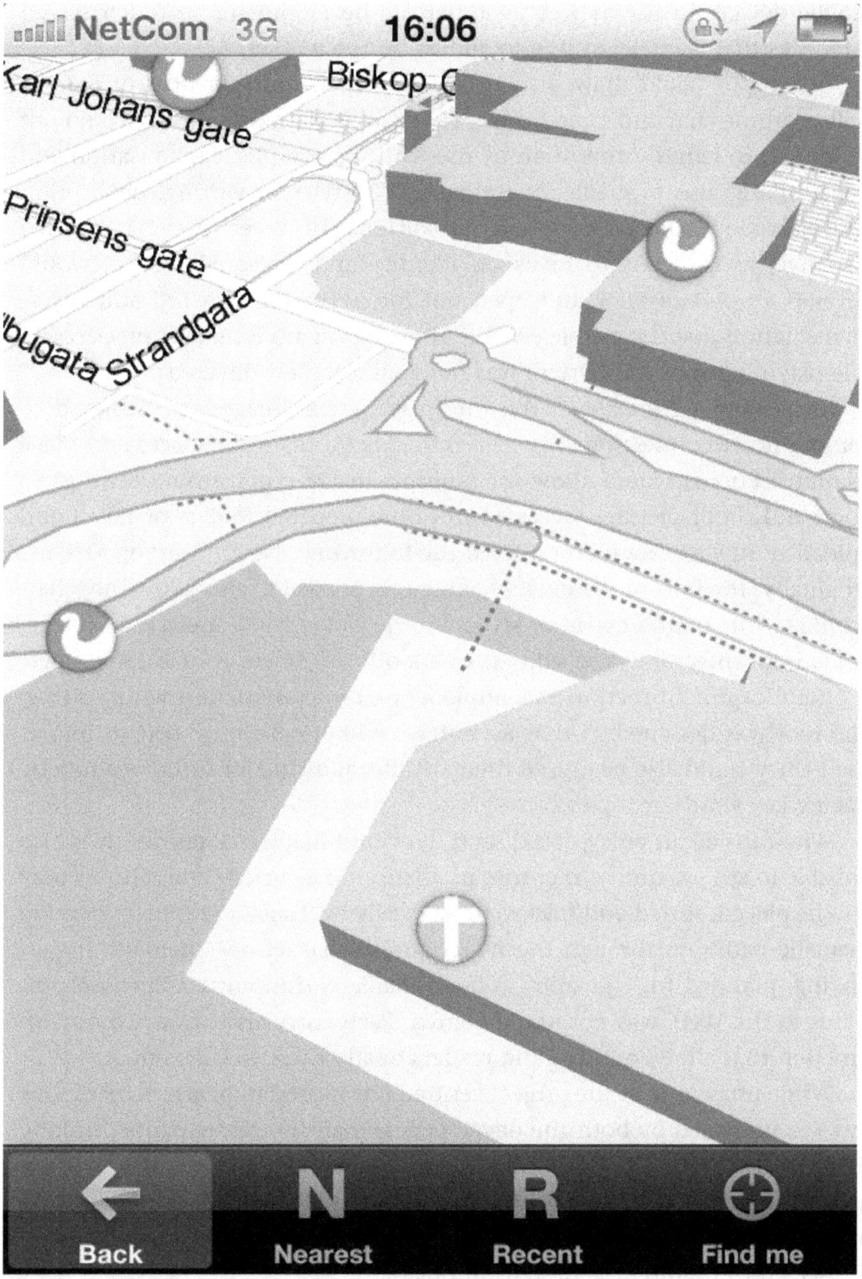

FIGURE 12.4 Map mode hybrid view showing volumes of main buildings

mobile devices in use as well as anticipate developments in touch screen interfaces that were beginning to appear on the market.

Text input could draw on existing traditions and practices of SMS as well as online chat and rapid forms of exchange. Narrative entries would not be limited in length other than by use, with paragraphs, capitalisation and hyperlinking also available. Narrative voice, styles of writing, description and dialogue would be the choice of authors. However, the character-led narrative, as a means to focus on identity and 'personal media', would develop a point of view and open out for stylisation of word and image in mediating a specific 'role'. While drawing on notions and practices of role-playing games, the system was not goal- or quest-directed.

In designing *NarraHand*, the interplay of text-image was designed to communicate connections between narrative voice, event spaces and place relations. Photos would allow for multiple image types, from drawings to point-and-shoot pictures, accessed from pre-prepared folders or taken and uploaded on site. The menu offered the following: 'New', 'Edit' or 'Delete'. Originally, the flow of the entry-making process would also allow the 'character' to write (with keypad or stylus on the Nokia N95), include an image, upload an entry and also edit it, or modify or delete it at a later stage. A photo 'capture' function was implemented in conjunction with writing. The tool was designed so that an entry could also be only text or image. Each entry could also be tagged thematically, allowing for future reading by tracing key words or topics.

When inside an entry, 'Next' and 'Previous' help track entries. It is also possible to see proximity of entries in relation to a current one. Entries may also be placed, sorted and tracked thematically by 'Labels'. This helps develop thematic pathways through the hypernarrative structure. Thematic hyperlinking, planned for the 'app' to be available within entries but also outwards to the Web, was not implemented. Each entry also has a 'Comment' function that allows authors and readers to add these to each entry.

While phase one of the project centred on the design of affordances, this too was informed by both the developers and author-participants thinking through their technical and communicative designs and characters while using the application during its development. Following principles and practices of collaborative and iterative design, we moved from a focus on the screen to trialling the designs in practice.

While phase one lasted longer than intended, its extension has allowed us to now adapt *NarraHand* for use with current touch phones with improved graphic and interface potential.

Conclusion

Much attention has been given to the discursive aspects of digital communication. Such communication occurs at the intersection of technological and semiotic affordances that are inscribed within digital hardware and software itself. As communicators, we construct and are shaped by these mediational technologies.

In this chapter we have argued that, along with more traditional concerns with form and meaning, scholars of discourse and creativity need also to consider the creative processes involved in the design of the mediational means that make communication possible and in many ways help to shape that communication. Engaging discursively, in and with design, in the development of semiotic resources is especially important in the context of the increased use of mobile digital technologies for communication. Locative media are still very much in development, and yet smartphone sales are booming worldwide. Attention is needed in the design of such media both as technological tools and as spaces for cultural expression.

The case of *NarraHand* presented here demonstrates how complex processes of co-design and collaboration are involved in the developmental shaping of resources for mobile narrative. Exploratory, iterative and participatory design approaches are all needed to tease out the discursive design aspects of multimodal communication (Morrison, 2010).

In terms of narrative discourse production, *NarraHand* took the conceptual designing of a framework for interaction into the dialogical development of a fully functioning mobile application. This tool allows f or subsequent creative expression in the second phase of the project now under way in which the designs presented above are being revised for use by a small authoring group of Africans living in Oslo writing and reading *NarraHand* on iPhones. Some of these authors were part of the iterative, collaborative design processes described above. Since its genesis, *NarraHand* has been more than a technical and narrative design. It has been part of an emerging discourse in the hands of authors who too have needed to find their way with mobile technologies oriented towards cultural expression:

Henry and Andrew meet at the central station to talk about their characters. They are now familiar with their smartphones and handwriting functionality. A few hundred metres away, the new Oslo Opera glistens in the summer sunlight.

Its ramps and rooftop open to the sky are swarming with visitors, ant-like on the white marble.

'It's taken me so long to land on a character', says Andrew.

'I've been thinking . . .' muses Henry. 'It's not just a question of taking on a role given in a game. It's like shaping your own avatar.' He looks over at the nearby café full of people. 'We have to be someone interesting for all these people, don't we? And the character has to be important to me'.

'So, have you landed on one?'

Henry laughs. 'It might not be what you expect!'

'That's the best thing about working with you', says Andrew.

'He's a . . . stand-up comic.' They now both laugh. 'And you?' asks Henry.

'I was talking to a friend of mine about a novel by Alain Mabanckou from the Congo with a porcupine as its narrator – and my character simply appeared. She's a rather small . . . white elephant.'

'Ha!' says Henry.

Notes

1 http://m4lit.wordpress.com

2 http://mrbellersneighborhood.com

3 http://wetellstories.co.uk/stories/week1/

References

Adami, E. and Kress, G. (2009) 'The social semiotics of convergent mobile devices: New forms of composition and the transformation of habitus' in *Multimodality: A social semiotic approach to contemporary communication*, Kress, G. (2009). London: Routledge, 184–97.

Arnall, T. and Martinussen, E. (2010) 'Depth of field: Discursive design research through film' Special Issue on Research by Design. *FORM Akademisk* **3** (1) Available: http://www.formakademisk.org/index.php/formakademisk/article/view/68

Benford, S., Crabtree, A., Flintham, M., Drozd, A., Anastasi, R., Paxton, M. and Tandavanitj, N. (2006) 'Can you see me now?'. *ACM Transactions on Computer Human Interaction* **13** (1) 100–33.

Bødker, S. and Bøgh Andersen, P. (2005) 'Complex mediation'. *Human-Computer Interaction* **20** 353–402.

Boyd, D. and Ellison, N. (2007) 'Social network sites: Definition, history, and scholarship'. *Journal of Computer-Mediated Communication* **13** (1) Available: http://jcmc.indiana.edu/vol13/issue1/boyd.ellison.html

de Lange, M. (2009) 'From always-on to always-there: Locative media as playful technologies' in *Digital cityscapes: Merging digital and urban playspaces*, de Souza e Solva, A. and Sutko, D. (eds). New York: Peter Lang, 55–70.

Diamantaki, K., Rizopoulos, C., Charitos, D. and Tsianos, N. (2009) 'LOCUNET: Location-based games as media communicating meaning' in *Digital cityscapes: Merging digital and urban playspaces*, de Souza e Solva, A. and Sutko, D. (eds). New York: Peter Lang, 148–167.

Gibson, J. (1979) *The ecological approach to visual perception*. Boston: Houghton Mifflin.

Hayles, C. (2008) *Electronic literature: New horizons for the literary*. Notre Dame, IN: University of Notre Dame Press.

Jenkins, H. (2006) *Convergence culture: Where old and new media collide*. New York: NYU Press.

Knutsen, J. and Morrison, A. (2010) 'Have you heard this? Designing mobile social software for discovery of independent music'. Special Issue on Research by Design, *FORM Akademisk* **3** (1) Available: http://www.formakademisk.org/index.php/formakademisk/article/view/66

Kress, G. (2000) 'Design and transformation'. In *Multiliteracies*, Cope, B. and Kalantzis, M. (eds), London: Routledge/Falmer, 153–161.

Kress, G. and van Leeuwen, T. (2001) *Multimodal discourse*. London: Arnold.

Løvlie, A. (2009) '*Textopia*: Designing a locative literary reader'. *Journal of Location Based Services* **3** (4) 249–76.

Mainsah, H. (2009) 'Cameroonians in Oslo, Diaspora and uses of the media'. *Nordicom Review* **30** (1) 83–94.

Manovich, L. (2009) 'The practice of everyday (media) life: From mass consumption to mass cultural production?'. *Critical Inquiry* **35** (2) 319–31.

Mäyrä, F. and Lankoski, P. (2009) 'Play in hybrid reality: Alternative approaches to game design'. In *Digital cityscapes: Merging digital and urban playspaces*, de Souza e Solva, A. and Sutko, D. (eds). New York: Peter Lang, 129–47.

Messeter, J. (2009) 'Place-specific computing: A place-centric perspective for digital designs'. *International Journal of Design* **3** (1) 29–41. Available: http://www.ijdesign.org/ojs/index.php/IJDesign/article/view/353/239

Moran, S. and John-Steiner, V. (2003) 'Creativity in the making: Vygotsky's contemporary contribution to the dialectic of development and creativity' in *Creativity and Development*, Sawyer, K., John-Steiner, V., Moran, S., Sternberg, R., Feldman, D., Nakamura, J. and Czikszentmihalyi, M. (eds). Oxford: Oxford University Press, 61–90.

Morrison, A. (2003) 'From oracy to electracies: Hypernarrative, place and multimodal discourses in learning' in *Digital media revisited*, Liestol, G., Morrison, A. and Rasmussen, T. (eds). Cambridge, MA: The MIT Press, 115–54.

Morrison, A. (2010) (ed.) *Inside multimodal composition*. Cresskill, NJ: Hampton Press.

Morrison, A., Westvang, E., and Skogsrud, S. (2010) 'Whisperings in the undergrowth: Performativity, collaborative design and online social networking' in *Exploring digital design*, Wagner I., Bratteteig, T. and Stuedahl, D. (eds). Vienna: Springer, 221–60.

Norris, S. and Jones, R. (2005) (eds) *Discourse in action*. London: Routledge.

Nova, N. and Girardin, F. (2009) 'Framing the issues for the design of location-based games' in *Digital cityscapes: Merging digital and urban playspaces*, de Souza e Solva, A. and Sutko, D. (eds). New York: Peter Lang, 168–86.

Page, R. (2010) *New perspectives on narrative and multimodality*. London: Routledge.

Rettberg, S. (2005) 'All together now: Collective narrative and collective knowledge communities in context'. Proceedings of *DAC 2005*, Copenhagen, 99–109.

Ryan, M.-L. (2006) *Avatars of story*. Minneapolis: University of Minnesota Press.

Ryan, M.-L. (2007) 'Narrative and the split condition of digital textuality' in *The aesthetics of net literature*, Gendolla, P. and Schäfer, J. (eds). Bielefeld: Transcript Verlag, 257–80.

Sawyer, K. (2006) *Explaining creativity: The science of human innovation*. New York: Oxford University Press.

Shotter, J. and Billig, M. (1998) 'A Bakhtinian psychology: From out of the heads of individuals and into the dialogues between them' in *Bakhtin and the human sciences*, Mayerfield Bell, M. and Gardiner, M. (eds). London: Sage, 13–29.

Urry, J. (2004). 'Connections'. *Environment Planning D: Social Space* **22** 27–37.

Vygotsky, L. (1962) *Thought and language*. Cambridge, MA: MIT Press.

Walton, M. (2009) *Mobile literacies and South African teens*. Cape Town: Centre for Film and Media Studies; University of Cape Town. Available: http://m4lit.files.wordpress.com/2010/03/m4lit_mobile_literacies_mwalton_20101.pdf

Wertsch, J. (1991) *Voices of the mind. A sociocultural approach to mediated action*. Cambridge, MA: Harvard University Press.

Acknowledgements

Our thanks go to InterMedia Lab, University of Oslo; the Inventio project (funded by the VERDIKT program Research Council of Norway, its members and annual seminar with Stanford University's Digital Humanities Lab), and colleagues at the Institute of Design, Oslo School of Architecture and Design (AHO). Special thanks to all participants in *NarraHand*. This research is an outcome of the gRIG Project (Guild for Reality Integrators and Generators) funded by the EU Culture 2000 Programme. On the *NarraHand* wiki, see: http://www.intermedia.uio.no/display/narrahand/Context

CHAPTER 13

Skating across timescales: Creativity in dynamical systems

Rodney H. Jones

Skating turned me into a whole new me.

Chun Jai, aged 16

In this chapter I will use principles from mediated discourse analysis and complexity theory to explore the relationship between creativity and learning in human ecosocial systems. Most of the discourse analytical approaches to creativity represented in this volume focus on the forms creative texts take, or the cognitive or social conditions of their production or interpretation. My focus in this chapter will be more on examining the kinds of social actions, social identities and forms of social organisation that are made possible by the creative texts that we make. In other words, my focus will be not so much on what makes a text creative as on why we make creative texts in the first place and what we do with them.

The data for this discussion come from a year-long ethnography of urban skateboarders in Hong Kong, which involved participant observation, in-depth interviews, and the collection of various media texts including a large number of amateur skateboarding videos which my participants shot, edited and regularly posted on YouTube, MySpace and various specialist websites for skateboarders. These videos attest to the fact that, for many skaters, skateboarding involves not just the mastery of complex physical skills, but also the mastery of the creative skills required to shoot and edit these videos, including an understanding of style, characterisation, narrative, rhythm and timing, and multimodal design, and an understanding of how to effectively combine images, music, fashion and architecture into complex, multilayered aesthetic products.

My primary emphasis will be on how these creative products aided my participants in learning how to skate and how to 'become skateboarders'. There has been considerable interest in the last thirty years in the relationship between learning and identity from a variety of theoretical perspectives including sociocultural theory (Bruner, 1996; Pollard, 1999), activity theory (Engeström, 1990; Sannino *et al.* (eds), 2009), discursive psychology (Harré, 1983), and literacy studies (Barton and Hamilton, 1998; Gee, 2003). 'All learning,' asserts Gee (2003: 51), 'requires identity work. It requires taking on a new identity and forming bridges from one's old identity to the new one.' This assertion points to an important aspect of identity when it comes to learning: its *emergent* quality. Learning is not a matter of identity 'transformation' in which one turns miraculously from one kind of person into another. It is an ongoing project of 'forming bridges' between momentary actions and interactions and longer developmental processes, between the self that I was before, who I am now, and the person who I aspire to be, and between my own identity project and the identity project of the group of which I am part. The role of creativity in the building of these bridges will be the main concern of this chapter.

Learning in dynamical systems

The view of learning described above is dramatically at odds with the traditional view of learning evident in most school curricula in which learning is seen as an essentially linear process following a hierarchy of skills in which lower-order skills such as remembering and understanding form the basis for higher-order abilities such as analysing, evaluating and, finally, creating. This view of learning is perhaps best exemplified by Bloom's Taxonomy (Bloom, 1956), in which learning is envisaged as a pyramid of objectives in which learning at the higher levels is dependent on one having attained prerequisite knowledge and skills at lower levels (Orlich *et al.*, 2004).

Work on learning influenced by systems theory (see for example Davis and Sumara, 1997) posits that, rather than a matter of linear, cause and effect processes, learning is 'occasioned' within complex dynamical systems of interrelations whose outcomes are difficult to predict, where multiple processes take place on different timescales, where analysing may precede understanding, and creating may precede remembering, and where tiny causes might have enormous effects, as in Lorenz's (1963) famous description of the flapping of a butterfly's wings in Brazil setting off a tornado in Texas. In such conditions knowledge and skills are insufficient for success. Barnett (2004: 254) suggests that successful learners within this model tend to be those who find ways to be 'self-energising and self-propelling' (p. 254). Students must be able to make connections between what they are doing now, frustrating and meaningless as it might seem, and what they want to do in the future.

One characteristic of dynamical systems is that they involve not just a single chain of actions or processes, but multiple processes taking place on multiple timescales. Lemke (2000b), for example, has described the multiple timescales on which school-based learning takes place, from the micro timescales of chemical synthesis, membrane processes and neural firings that make up brain activity to the longer timescales of utterances which take place within exchanges, which take place within lessons, which take place within school days, which take place within semesters, years, lifetimes, and larger cultural and natural processes. Each action students perform in classrooms not only participates in local constructions of meaning on shorter timescales, but also 'takes part in the systematic networks of interdependent activities that sustain careers, institutions and societies over much longer timescales' (Lemke, 2001: 84).

A similar mapping can be made of the multiple processes implicated in the act of skateboarding, illustrated in Figure 13.1, from the tiny muscular movements that go into making a move, to the rapid series of moves that constitute a trick, to the sequences of tricks that make up what are called lines, to the skating session, the day, the week, the year, the career of a particular skater, the lifespan of an individual, the 20-odd-year history of skateboarding in Hong Kong where I did my research, the approximately 50-year-history of global skateboarding and longer cycles of social, economic and technological change. In a single moment of skating all of these timescales are potentially present. The way they are experienced depends on our perspective. A particular trick, for example, can be seen as made up of various micro-moves or as a moment in the learning trajectory of a particular skater, or in terms of the evolution of the sport of skateboarding.

The point of all of this is that, whether one is a student in a science class aspiring to be Albert Einstein or a skater in a skate park aspiring to be Tony Hawk, success depends chiefly on one's ability to play with these different perspectives within complex systems of multiple processes, and to find some way to make connections between information generated on one timescale and information generated on another.

The key to understanding how this happens, and the thing that separates human ecosocial systems from natural systems is what Lemke (2000a, 2000b, 2001) calls 'heterochrony'. In natural ecosystems processes on radically different timescales are buffered from each other by the intervening timescales

Timescale	Examples
History of Social and Technological Change	• Human history
Global History of Skating (approx 50 yrs)	• Histories of global brands/global flows of culture; Global political and economic histories
Local History of Skating (approx 20 yrs)	• Local political and economic histories; Flows of people (expats/'backflow')
Life Span (approx 80 yrs)	• Birth, old age and death; Generations
Skateboarding Career (up to 15 yrs)	• Academic and work careers; Rates of bodily healing; Coming and going of skate spots
Year/Seasons (365 days)	• Annual skating competitions; Lifespan of equipment; Product releases; School holidays
Week (7 days)	• Working week; School week
Day (24 hrs.)	• Circadian rhythms; Rotation of Earth; School and work schedules
Session (approx. 1–8 hrs.)	• Skate park opening hours; School schedules; Security guard patrol schedules; Meals
Heat (approx 10–30 minutes)	• Limitations on physical stamina (entropic cycles); Conversations; Skate videos
Line (approx 5 sec-several minutes)	• Musical verses; Conversational exchanges
Trick (approx 1–5 seconds)	• Respiration; Heartbeats; Orgasm; Ocean waves; Words
Move (less than a second)	• Neural processing; Muscle fiber discharge; Rapid Heart rate; Surface vibrations

FIGURE 13.1 Timescales of skateboarding

so that information cannot be readily exchanged. Lemke (2001: 82) gives two examples: 'climate change and local weather tend to be highly buffered from one another (and) the long-term ecological succession in a large area of forest is not determined by the cutting of one tree, and vice versa'. This buffering is one reason why natural ecosystems evolve so slowly. In *social* ecosystems, on the other hand, processes on non-adjacent timescales are able to interact in such a way that events remote in time (such as, say, getting a job as a chemical engineer) may be more relevant to what a student is doing in the classroom at a particular moment than events nearer in time (his next class, or his dinner, or his date with his girlfriend).

What makes this possible, according to Lemke, and what makes human learning different from climate change or biological evolution, is the presence of material semiotic artifacts in human societies. Semiotic artifacts such as textbooks, assignments, examinations, transcripts, diplomas and resumes, for example, enable students in chemistry classes to make connections between what they are doing today and what they did last week, and between shorter-scale classroom tasks and longer-scale career trajectories. In describing this phenomenon, Lemke (2001: 83) quotes Serres and Latour (1995), who used the metaphor of 'folded time' in which 'events remote in time from one another may be culturally more relevant to a present event than other events nearer in time'.

What I would like to explore here, then, is the role creative texts play in human learning and identity formation through their ability to mediate the multiple bodily, social, institutional and historical processes that occur on different timescales. Put more simply, my goal is, following Lemke, to understand how moments add up to lives (2000b: 273), or, more specifically, how isolated tricks in skate parks, alleys, abandoned lots and pedestrian plazas add up to the identities of urban skateboarders.

Skateboarders and digital video

The semiotic artifact I am most interested in is the amateur skate video. From the early days of skating, video has always played an important role in the sport. Before the internet, skaters used to trade Betamax and VHS tapes of themselves and gather in skate shops to watch grainy videos over cigarettes and beer. Nowadays they post their videos on YouTube, MySpace and Facebook, sending these documents of momentary tricks done in specific spots across longer timescales and spatially distant places. Learning how to shoot, perform in, and edit video is to some extent part of learning to be a skater. Skaters regularly bring video equipment with them when they skate to document their activities, and spend hours meticulously editing these videos and setting them to music. Videos are part of every stage of the skater's career from the beginning, when he[1] carefully studies them to understand how to do a particular trick or to inspire himself, to later stages where he makes his own videos, maybe sending them to shoe and clothing companies in the hopes of being sponsored.

For many of my participants, the processes of watching, shooting and editing digital video played a central role both in their decision to become skaters and in their ability to sustain their interest in the sport and their ties with other members of the community. One participant related how he first became attracted to skateboarding, saying, 'When I was little I went to my cousin's home; he used to play some videos and I was like "wow, how could their feet stick to the board? How do they do this?" And when I got the chance I wanted to try to do it myself.' (Simon, 19) Another described the role of video in his skating career this way:

> When I first started skating my friend had a camera and we started to film some short videos. The camera was very low-tech. The quality of the clips was really bad and shaky. But we couldn't wait to add some songs to it. Later we finally got a DV camera. We were determined to buy a DV camera no matter how poor we were. Then we started to put our clips up online. Sometimes you feel so happy when people compliment your work. And if it sucked, then next time we would do better.
>
> (On Jai, 20)

While self-documentation has always been important for skateboarders – dating back to the 1970s and 1980s when they took still photos of each other and distributed them in underground 'zines' (Weyland, 2002) – the advent of digital video and the internet has had a dramatic impact on the ability of skaters not just to produce and distribute images, but to manipulate and alter them in creative ways. New technologies allow any teenager to shoot and edit digital video to a level of sophistication, which, less than a decade ago, required professional knowledge and expensive equipment.

The skate video is not a random collection of shots of people skating. It is a genre with clear conventions, which have particular meaning and currency within this discourse community. Skate videos typically consist of footage of skaters executing a series of 'lines' across a variety of urban landscapes artfully edited and set to music. The combination of editing, music, movement, fashion, and interaction between the skaters and other people in these environments (passers-by, police officers) often creates a kind of narrative of a particular skater or a particular 'crew'.

These videos serve multiple functions for skaters. For individuals, they serve as inspirational and instructional tools and as proof of their accomplishments, texts of self-promotion that they might submit to skate shops or equipment manufacturers to seek sponsorship. For groups of skaters, they serve to create social cohesion and to communicate that cohesion to the community as a whole, and for the community they preserve the lore of lost skate spots and legendary members. At all of these levels – the individual, the crew, and the community as a whole – they engender what dynamical system theorists call *emergence*, the arising of new properties, behaviours, abilities and identities out of the interaction of all the various processes within the system occurring on different timescales. They do this, I will argue, by helping skaters integrate short-scale processes of tiny muscular movements with longer-scale processes occurring over days, weeks, years, individual skateboarding careers and the history of the community. They help skateboarders turn moments into lifetimes.

The move

The shortest timescale accessible to the consciousness of the skater is that on which the move occurs. A move consists of a rapid series of contractions, which cause variations of weight distribution on the board or against the ground. Moves are often so slight, so subtle, that in the actual practice of skating they are felt rather than seen. It is on this timescale that the relationship between time and the physical materialities of space is most apparent. Skaters skate on physical surfaces, and different surfaces have different speeds. They experience the city as felt vibrations as their boards roll and grind on different surfaces, vibrations that they learn to read in ways others cannot. At the same time skaters become part of these spatial/temporal textures themselves, altering the surfaces that they wax and grind so that the actions of the moment are linked to longer timescales of building construction and repair and the rise and fall of skate spots, a linkage which for skaters is imbued with a cruel irony: the more you grind a surface the better it is for skating but also the greater the chances that these surfaces will become damaged, inciting authorities to take steps to prevent skaters from using them.

In the context of video, these physical sensations become visible, especially as clips are isolated and played at various speeds. When skaters watch any skating video they watch it differently from non-skaters, paying attention to tiny microsecond processes that are invisible to the untrained eye. As one of my participants put it:

> It's like you can see everything you did. Maybe when you did it it felt all right, but when you play it back you can see where you're sketchy, or maybe you're moving your arms around like a dumbass. When you watch the times you didn't land it, you can see exactly where you made the mistake.
>
> (Gary, 21)

The process of learning how to watch a skate video in a way in which moves are as legible as letters in a printed word plays an integral part in the education of the skater, linking the momentary microseconds of the move to longer timescales of learning trajectories. Beek and his colleagues (2003), applying dynamical systems theory to the mastery of athletic techniques, argues that one of the most important aspects of expertise in sports is the ability to strategically focus one's attention on processes occurring on shorter and then longer timescales as the learning process unfolds, and this attention to the move is the first step in developing this ability.

The trick

Whereas the move is the smallest 'feelable' unit, the trick is the smallest nameable one, and nearly all of them have names like the ollie and the nollie and the nose grind and the half-cab flip. The trick is really the whole point of skating, the moment where, in the midst of the body's micro-measurement of space and time and texture, the skater is lifted out of time. The biological process to which my participants insisted the trick might best be compared is the orgasm – 'it lasts about the same amount of time', said one, 'and makes you feel about the same' (Gary, 21). Tricks, however, are not so easy to execute. They are moments in which hours and weeks and months and even years of repetitive practice converge, moments in which the whole history of skateboarding is collapsed, the labels that they are given linking skaters of today with skaters of the past. The half-cab flip, for example, is named after American skater Steve Cabalerro who invented it in 1981.

The trick is also the central unit of the skate video. In fact, the whole point of the skate video is to capture that elusive and transitory moment into a permanent, perpetually re-playable document that proves that 'you did it'. In videos showing more daring tricks, what often happens in the editing is a stretching out of these moments, a slowing-down of time to reveal the beauty and intricacy of the actions. These slow-motion segments in skate videos, however, are not just about making the trick look beautiful, but also about mediating the 'objective' time of the stationary observer with the relative psychological time of the skater in motion. 'That's really the way it feels when you're doing it', one of my informants insisted, 'like time is expanded, like you're outside of time' (Owen, 20).

The line

The next scale up is that on which the 'line' occurs. A line is a series of one or more tricks done in succession along a trajectory of surfaces and obstacles. It is in the process of doing lines that the skater rewrites the city, carving new pathways through its terrain. It is a re-sequencing of the urban landscape, not just as a series of tricks, but as a series of relationships with the material world and a way of re-ordering those relationships in time. It constitutes the linking together of the rhythms of the different textures of the surfaces upon which tricks are performed, creating what Lefebvre (2004: 20) calls a 'garland of rhythms'.

If tricks are difficult, however, successful lines involving multiple tricks are even more improbable. The flawless lines portrayed in skate videos, in fact, rarely happen – what generally happens is a lot of falling down. Often it takes hours of footage for skaters to assemble enough successful tricks to string together into just a few minutes of finished video. Here is where the editing capabilities of digital video become particularly important. 'I'm really not that good, you know', one participant admitted, 'but if I'm good at editing, I can make myself look like a pro' (On Jai, 20).

The lines documented in skating videos, then, and the chains of lines that give the impression of the skater seamlessly travelling through the urban landscape, portray skating not as it is, but as it ought to be. These videos are at once documents of serendipitous moments and the compression of many hours and days of failed attempts, at once documents of what really happened, and idealised versions of what might happen, or could happen or should happen, produced through careful selection and editing. This is where these artifacts function in particularly powerful ways for skaters, allowing them to string together their successes into idealised portrayals that reveal not just their past glory but also their future potential.

Along these three levels that I have described so far, we can see operating what Lemke (2000a, 2000b, 2001) calls the 'principle of alteration', the tendency for processes on adjacent timescales to alternate between continuous variation and categorical information. According to the principle of alternation, as we move from level to level, information from lower levels is reorganised in such a way that differences of degree below matter as differences of kind above, and vice versa. For example, the differences in degree that go into making speech sounds are transformed at the level of the word into typological meaning (differences of kind), to be transformed on the level of the utterance, with its variations of stress, volume and intonation to more topological meaning (differences in degree). In skating, the topological experience of height, distance, tactility and transition associated with the move is, on the level of the trick, experienced as a discrete set of types, to again shift on the next level to the more continuous experience of the line.

The session

These tricks and lines are practised and perfected in the context of sessions, which typically last several hours and have their own rhythms and spatial-temporal organisation, with skaters often moving in cyclical formations back and forth across obstacles. These rhythms also take on a spatial organisation linked to longer timescales of the skater's learning trajectories, with younger, more inexperienced skaters occupying certain spaces and attempting certain obstacles and more experienced skaters occupying others.

These sessions take on still different spatial-temporal organisation when they involve skating street spots rather than in skate parks (a practice universally preferred by all the skaters that I've met). Street sessions almost always involve migrating from spot to spot as police or security guards chase crews of skaters away. In a single night a crew might visit four or five spots throughout the city. And so street skating sessions are entrained in unique ways to circadian cycles and the schedules of other residents of the city. In Hong Kong, for example, street skating is almost always done at night, partly because the streets are too crowded with pedestrians during the day and partly because authorities are less likely to tolerate such breaches of public order in broad daylight. Another reason is that, unlike many other cities, skating is more possible in Hong Kong at night because of the brightness of the city lights, and this has become part of the unique character of the Hong Kong skating scene where skaters sometimes refer to themselves as 'children of the night'.

Individual careers

Over time, as sessions follow still more sessions, and as traces of these past moments accumulate in the form of old video footage, skaters develop a sense of both their technical progress and the progression of their status within the community. One of the most important functions of skate videos is to serve as a visual record of a skater's learning trajectory, allowing him to understand how he has improved and what he still needs to work on. This record encourages skaters to view their learning from a broader temporal perspective in which the failures and frustrations of the moment are subsumed into longer narratives of progress.

This progress, however, is as much about identity as it is about technique: the videos are records of the evolution of their 'selves' and their relationships, of their tastes in fashion and music, of their mastery of individual styles, of various repertoires of movement and expression and attitude and interactional routines, of their cultivation of 'cool'. One participant described the process like this:

> Everybody films these videos and shows them to everyone around, asking, 'Am I getting better? Am I good?' . . . For the music we pick the songs we love. It's like giving a theme song to follow your own rhythm, your own skill, your own speed. And the music can create a perfect match. I can review my skateboarding skills from the first tape to the last. If I keep filming, there must be some improvement. I mean, I look at myself in those old tapes, and it's sick. Even what I'm wearing. It's like I'm a totally different person now.'
>
> (Johnny, 17)

It is central to the ideology of skating that learning is – or should be – a totalising and utterly transformative experience, not just a matter of being able to do something, but a matter of being able to be someone within a social group, and these transformative experiences rub up against the rhythms of other careers taking place on the same timescale, in particular, academic careers, in which their experience of learning is rarely so transformative, and the transformative experiences they have skating are rarely valued.

Local histories/global histories/cultural flows

These nocturnal migrations from spot to spot that are stitched together to form the careers of individual skaters, memorialised in hours of old footage stored on hard drives, are replicated on the longer timescale of the local history of skating, which is essentially a history of disappearing space. Crews find skateable spots, these spots become popular, and then the authorities move in to close them down with regular security patrols or the installation of physical barriers that make skating impossible. Again, time and space are intertwined as historical periods take on the names of the spots popular in those times and subsequently lost – the Shatin era, the Cultural Center era, the Immigration Tower era. And so these videos link moves and tricks and lines to longer-timescale processes of building construction and repair and the rise and fall of skate spots.

Skateboarders have an almost mythological relationship with place, and the videos they make are not just documentaries of individual skaters or crews, but celebrations of the places where they skate and have skated in the past. This attention to place and culture also helps to link local skateboarding to longer global histories of skateboarding which share this phenomenon of the transitory nature of space – lost local spaces take their place within a catalogue of legendary spots like the Embarcadero Center in San Francisco and Philadelphia's Love Park, creating for local skaters a sense of solidarity with other skaters and giving to the activity the feeling that it is not just a sport, but a movement.

Related to this sense are the longer-timescale processes of global flows of cultural products, of fashion and music and language and attitude, flows that are entrained not just with global economic cycles but also with the cycles of school holidays and the travel of individual skaters who carry these cultural products in and out of Hong Kong, introducing them to their peers.

In communities of Hong Kong skateboarders these global flows of culture, discourse and language are a matter of everyday life. One thing that characterises these communities is variety in terms of age, ethnicity and place of residence. Most of the skateboarders I met in Hong Kong speak multiple languages, participate in multiple cultures and have lives that straddle multiple localities. They might live part of the year in Hong Kong but go to school in the UK or USA, or might have lived parts of their lives in different places. They represent a group of young people who are not just products of globalisation but agents in it as they migrate among multiple cultural milieux, acting as linguistic and cultural brokers carrying stories, music, fashions and other cultural products across timescales and places. 'Years ago', one participant recalled, 'there used to not be a punk fashion in skateboarding in Hong Kong. But then this guy moved back from the UK, and he really looked like the guy from the Sex Pistols. And then after that people started to wear tight and all that. Then there was like a punk crew' (Gary, 21).

This sense of mediating not just physical environments but also cultural environments is also an important aspect of Hong Kong skateboarding videos, which mix together English, Cantonese and Mandarin music, the landmarks of different locales – Beijing, Hong Kong, Sydney, Bangkok – and the pastiche of languages and speaking styles that locals and visitors share and imitate, appropriate and adapt.

Lifetimes

These careers become part of lifetimes that stretch out before these young skaters in improbable imaginings that exasperate their parents and teachers. Every skater I talked to insisted that they would skate forever, ignoring the physical reality that after a certain age the body simply cannot tolerate falling down so much and the economic reality that almost nobody can make a living from skating. The idealised present folded into those moments captured on video stretches out into an idealised future of perpetual youth and endless challenges. As one participant described it: 'It's like a history book of your life and your friends' lives. It's like, it's always going to be there. It'll always be there' (Jason, 18).

Finally, these videos serve as links between momentary everyday actions and imagined futures. The videos themselves might even be tickets to these futures, auditions for sponsorship by DC or Nike. But usually not. And it does not really matter. What matters is that they help skaters to imagine what might be possible by bringing together processes on all of the different timescales of their lives from the level of the move, to the trick, to the line, to the session, to the career, to the various local and global histories of skateboarding, fashion and popular music, the rhythms of these timescales carefully synchronised so that the sounds of the skateboard along the surface of the ground are entrained with the beats of this skater's favourite song, to trajectories of learning that have brought this skater to this moment. Skate videos are both tools for learning and documents of the learning process. As members of crews work together to film one another and to edit their footage, the videos serve as a focus for group action and group identity which is reinforced when the videos are distributed and others in the community recognise a particular group of skaters as a crew. They help individuals and groups craft for themselves social and cultural identities.

But, most importantly, they serve not just as retrospectives of past accomplishments and histories but also as anticipatory discourse, portrayals of a mythical world of skating where tricks are always landed, where perfect lines extend across disparate geographies of legendary landscapes which may have long since been reappropriated by authorities to become part of the stories and lore of the community. It is the ability to imagine this world that fuels these young people's progress, that propels them forward with the self-generated energy that Barnett (2004) observed in successful learners. As one participant put it:

> It's a way to like push yourself. You may try something down something not so hard, but if it's being filmed you want to try something even harder, to push yourself, to bring yourself to the next level.
>
> (Jason, 18)

Conclusion

So what can these videos and similar kinds of creative texts teach us about learning? One thing they demonstrate is that learning does not have to follow the linear progression of Bloom's Taxonomy, it can also be approached as a series of non-linear recursive processes and feedback loops in which application and synthesis might precede understanding, and creating is not necessarily an *outcome* of discovery, but might be a first step towards it. We learn because we can imagine. Central to this process are the opportunities these creative texts provide for skaters to make meaning on multiple scales of human experience, linking moment to moment and event to event, person to person, activity to activity, and to imagine idealised futures through a process of editing together past successes.

Therefore the simple conclusion I can come to as to the role creativity plays in our lives, at least based on this example, is that it enables us to 'fold time', and being able to fold time is an enormously powerful ability. These poems, pictures, skateboarding videos, web pages, and musical compositions that we make, and all of the other kinds of creative discourse analysed in the pages of this book are not ends in themselves; they are tools we use in the real creative act, the act of creating ourselves. They are where we consecrate our memories and imagine our destinies. They allow us to make some kind of sense of where we are going and where we have been. They are the flapping of the butterfly's wings which set off chain reactions in the dynamical systems of our lives and sometimes, if we are lucky, take us to the next level.

Note

[1] Throughout this chapter I use the male personal pronoun to refer to my participants and the skateboarders in general, simply because all of my participants and a vast majority of skateboarders are male. That is not, however, to say that female skateboarders do not exist, and the trials and tribulations they experience in positioning themselves within this predominantly male sport have been documented by, among others, Becky Beal (Beal, 1996; Beal and Wilson, 2004).

References

Barnett, R. (2004) 'Learning for an unknown future'. *Higher Education Research and Development* **23** (3) 247–260.

Barton, D. and Hamilton, M. (1998) *Local literacies: Reading and writing in one community*. London: Routledge.

Beal, B. (1996) 'Alternative masculinity and its effects on gender relations in the subculture of skateboarding'. *Journal of Sport Behaviour* **19** (3) 204–20.

Beal, B. and Wilson, C. (2004) 'Chicks dig scars': Commercialisation and transformations of skateboarders' identities' in *Understanding lifestyle sports: Consumption, identity and difference*, Wheaton, B. (ed.). Oxford and New York: Routledge, 31–54.

Beek, P. J., Jacobs, D. M., Daffertshofer, A. and Huys, R. (2003) 'Expert performance in sport: Views from the joint perspectives of ecological psychology and dynamical systems theory' in *Expert performance in sports: Advances in research on sport expertise*, Starkes, J. L. and Ericsson, K. A. (eds). Champaign, IL: Human Kinestics, 321–44.

Bloom, B. S. (1956) *Taxonomy of educational objectives: The classification of educational goals*. Chicago: Susan Fauer Company.

Bruner, J. (1996) *The culture of education*. Cambridge, MA: Harvard University Press.

Davis, B. and Sumara, D. (1997) 'Cognition, complexity and teacher education'. *Harvard Educational Review* **67** (1) 105–25.

Engeström, Y. (1990) *Learning, working and imagining: Twelve studies in activity theory*. Helsinki: Orienta-Konsultit.

Gee, J. P. (2003) *What video games have to teach us about learning and literacy*. New York: Palgrave MacMillan.

Harré, R. (1983) *Personal being*. Oxford: Blackwell.

Lefebvre, H. (2004) *Rhythmanalysis: Space, time and everyday life* (trans. Elden, S. and Moore, G.). London: Continuum.

Lemke, J. L. (2000a) 'Opening up closure: Semiotics across scales' in *Closure: Emergent organizations and their dynamics* (Volume 901: Annals of the NYAS), Chandler, J. and van de Vijver, G. (eds). New York: New York Academy of Science Press, 100–11.

Lemke, J. (2000b) 'Across the scales of time'. *Mind, Culture and Activity* **7** (4), 273–90.

Lemke, J. (2001) 'Discursive technologies and the social organization of Meaning'. *Folia Linguistica* **35** (1–2) 79–96.

Lorenz, E. N. (1963) 'Deterministic nonperiodic flow'. *Journal of the Atmospheric Sciences* **20** (2) 130–41.

Orlich, C., Harder, R., Callahan, R., Trevisian, M. and Brown, A. (2004) *Teaching strategies: A guide to effective instruction*, (7th edn), Boston: Houghton Mifflin Company.

Pollard, A. (1999) 'Towards a new perspective on children's learning?'. *Education* 3–13 **27** (3) 56–60.

Sannino, A., Daniels, H. and Gutiérrez, K. D. (eds) (2009) *Learning and expanding with activity theory*. Cambridge: Cambridge University Press.

Serres, M. and Latour, B. (1995) *Conversations on science, culture, and time.* Ann Arbor: University of Michigan Press.

Weyland, J. (2002) *The answer is never: A skateboarder's history of the world.* New York: Grove Press.

Index

www.ingramcontent.com/pod-product-compliance
Lightning Source LLC
LaVergne TN
LVHW011122300926
849160LV00008B/482
9781408251881